T0230859

# Essential Statistics for Medical Practice

## A case-study approach

### *D.G. Rees*

Senior Lecturer in Statistics
Oxford Brookes University
UK

**CRC Press**
Taylor & Francis Group
Boca Raton London New York

CRC Press is an imprint of the
Taylor & Francis Group, an **informa** business

First published 1994 by CRC Press
Taylor & Francis Group
6000 Broken Sound Parkway NW, Suite 300
Boca Raton, FL 33487-2742

Reissued 2018 by CRC Press

© 1994 by D. G. Rees
CRC Press is an imprint of Taylor & Francis Group, an Informa business

No claim to original U.S. Government works

ISBN 13: 978-1-315-89280-1 (hbk)
ISBN 13: 978-1-351-07190-1 (ebk)

Typeset in 10/12 pt Palatine by Best-set Typesetter Ltd., Hong Kong

Visit the Taylor & Francis Web site at http://www.taylorandfrancis.com and the
CRC Press Web site at http://www.crcpress.com

*To the memory of Florence Nightingale, her work and her inspiration*

*With Miss Nightingale statistics were a passion and not merely a hobby . . . But she loved statistics not for their own sake, but for their practical uses. It was by the statistical method that she had driven home the lessons of the Crimean hospitals. It was the study of statistics that had opened her eyes to the preventable mortality among the Army at home . . . She was in very serious, and even in bitter, earnest a* passionate statistician.

# Contents

# Preface

Books on medical statistics usually start with elementary statistical techniques of data presentation and analysis, building up to more advanced techniques. Relevant examples from the literature, for example medical journals, are quoted as each technique is introduced.

This book is different! I start from where the medical professional starts when reading the literature, namely with a medical investigation which nearly always contains some use of statistical methods. So in Part One I have taken six real case studies from recent medical journals and discussed each in turn, using a common format, with particular emphasis on the statistics. Forward references are made from the case studies to examples in Part Two, which draws together similar examples and discusses the background assumptions, limitations and applicability of the statistical methods described.

I have limited this book to only the most basic, but nevertheless useful, methods, because my aim is to provide a readable and relatively short primer, rather than a long text which might deter the busy health professional.

# Acknowledgements

*Miss Nightingale's mastery of the art of marshalling facts to logical conclusions was recognised by her election in 1858 as a member of the Statistical Society.*

I would like to thank the editor of the *British Journal of General Practice* (formerly the *Journal of the Royal College of General Practitioners*) for permission to reproduce five articles (case studies 1–5), and the production director of the *British Medical Journal* for permission to reproduce one article (case study 6).

I would further like to thank the following authors and publishers for their kind permission to adapt from the following tables: Pearson, E.S. and Hartley, H.O. (1966), *Biometrika Tables for Statisticians, vol I, 3rd edition*, Cambridge University Press, Cambridge (Tables B.1–B.4); and Mead, R. and Curnow, R.N. (1983), *Statistical Methods in Agriculture and Experimental Biology*, Chapman & Hall, London (Table B.5).

The quotations before or after various sections and chapters are taken from a biography written by Sir Edward Cook (1913), *The Life of Florence Nightingale*, Vol I 1820–1861 and Vol II 1862–1910, Macmillan, London, with permission from the publishers. The quotations are the biographer's words unless otherwise stated.

Thanks also to David Mant, Department of Community Medicine and General Practice, University of Oxford, and to Peter Anderson, formerly Director of HEA Primary Health Care Unit, Churchill Hospital, Oxford, for initial discussions and assistance with the choice of relevant case studies.

Most of all, I am grateful to my wife, Merilyn, for her help in the production of the various drafts and for her support and encouragement throughout.

# Overview of the use of basic statistical methods in medical studies

Data are collected from patients in order to diagnose and treat their medical conditions, and to make them well again. In medical studies we usually need to collect data from one or more groups of patients who have something in common, such as suffering from the same disease. We may then wish to compare sub-groups of patients, for example those treated by one method and those treated by another method. The simplest way to compare the sub-groups will probably be in terms of means or percentages, depending on the type of variable of interest. For example, if the treatments given to sub-groups 1 and 2 are one of two drugs designed to reduce blood pressure, we would wish to compare the mean reduction in blood pressure for each of the sub-groups. Whereas, if the two treatments given result in either 'success' (disease cured) or 'failure' (disease not cured) for each patient, we would wish to compare the percentages achieving success for the treatments. Means and percentages are examples of summary (or 'descriptive') statistics, and these and others are discussed in the first chapter of Part Two (Chapter 7).

The other aspect of the medical data we collect is that they represent only some of the possible data that we could have collected. We, of course, restrict ourselves to collecting only data for variables which are relevant to the objectives of our study. Even so, if our objective is, for example, to 'compare the reduction in blood pressure resulting from the applications of drugs A and B' we will have to restrict ourselves to collecting data from *some* of the patients taking these drugs for this purpose. Statisticians say that we take 'samples' from 'populations'. In the example stated the statistician envisages two populations. One consists of all the blood-pressure data for all patients taking drug A, the other similarly for drug B. He/she also envisages two samples, one from each population. In the example, these samples consist of the blood-pressure data for the patients we actually include in our study. The statistician regards the 'sample data' as 'known', in the sense that he can calculate various statistics from them. He also regards the 'parameters' of the population as 'unknown', for example he cannot calculate the mean reduction in blood pressure for the population. However, he can *estimate* population parameters from sample data and express his results in terms of confidence intervals. Alternatively, he can specify a hypothesis about a population parameter, and test whether the sample data support the hypothesis or not.

Statisticians refer to the subject of drawing conclusions about populations from sample data in terms of confidence intervals or hypothesis tests as 'statistical inference'. Chapters 8, 9, 10 and 11 of this book cover those inferential methods concerning means and percentages which are most commonly used in medical studies.

While we may wish to compare two groups of patients in terms of means of one particular variable, it is sometimes useful to study the way in which two (or more) variables are related. To what extent, say, can one variable be used to predict another variable? For example, it is possible to predict the basal metabolic rate (BMR) of an individual from his/her body weight (given also the sex and age group of the individual) using a simple linear equation. This is a useful practical idea because body weight is much easier to measure than BMR. Statisticians call such an equation a 'simple linear regression equation', and the ideas of 'regression analysis' can be extended to cases where we might usefully employ more than one variable to predict another (so-called 'multiple regression analysis'). An idea related to regression is called 'correlation'. The degree to which two variables are linearly related can be measured in terms of a 'correlation coefficient'. Chapter 12 of this book is an introduction to regression and correlation, and includes the concepts of statistical inference mentioned above.

Chapter 13 in one sense is unrelated to previous chapters. It deals with the value of diagnostic tests in determining the true condition of a patient – in terms of the sensitivity and specificity of the tests. (The sense in which they *are* related is in terms of drawing conclusions in the face of uncertainty, which is inherent in all statistical inference.)

Chapter 14 deals with the important topics of the various types of medical study and the size of such studies. Some might suggest that this chapter should have come earlier in the book. However, the concepts required to decide study size using a 'scientific' method are those of statistical inference which runs through almost the whole of the rest of the book!

*Part One*

# Discussion of Case Studies

# Preliminary trial of the effect of general practice based nutritional advice

*In April 1855 I [FN] undertook this Hospital, and from that
time to this (November 1855) we cooked all the Extra Diet for
500 to 600 patients, and the Whole Diet for all the wounded
officers by ourselves in a shed. But I could not get an Extra Diet
Kitchen till I came to do it myself. During the whole of this time
every egg, every bit of butter, jelly, ale and Eau de Cologne
which the sick officers have had has been provided out of Mrs
Samuel Smith's or my private pocket. On November 4 I opened
my Extra Diet Kitchen.*

Case study 1 is an article by John A. Baron, Ray Gleason,
Bernadette Crowe and J.I. Mann taken from the *British
Journal of General Practitioners*, **40** (1990), 137–41.

SUMMARY. Despite formal recommendations for dietary
change to reduce the incidence of ischaemic heart disease, the
acceptability and effectiveness of the proposed diets have not
been well investigated in population based studies. In this
preliminary investigation of nutritional advice in a well population, subjects in one group practice were randomized to
receive either dietary instruction or simple follow up without
instruction. The dietary recommendations were well received,
and a substantial proportion of subjects reported altering their
diets in accordance with them. There were modest beneficial
changes in plasma lipid levels among men. Thus, using general

practice as an avenue for promoting dietary change is feasible, and may be effective among men.

## INTRODUCTION

Official organizations in the United Kingdom have recommended that the British public modify its present diet in order to reduce the incidence of ischaemic heart disease.[1,2] Common to many of these recommendations is advice concerning maintenance of optimal body weight, increased dietary fibre, reduced total fat intake, and an increased ratio of polyunsaturated to saturated fat intake. Although there have been several clinical trials that have studied dietary intervention for ischaemic heart disease, these have stressed multifactorial intervention (with a variable dietary focus), have featured diets high in total fat, or have used a very high ratio of polyunsaturated to saturated fats.[3] Only a few of the studies have been population based.[3] The acceptability of the currently recommended dietary advice to the healthy UK population (as distinct from patients or high risk subgroups) has not, therefore, been well studied. Also, for the general population, there is little information concerning the effect of such dietary change on the metabolic parameters that are associated with ischaemic heart disease, such as serum lipoproteins.

This report describes the results of a preliminary, general practice based randomized controlled trial of the current dietary recommendations. Its aim was to assess the acceptance of the diet to a healthy UK population, to ascertain whether a general practice based approach would promote its use, and to provide preliminary information on its effectiveness in lowering lipid levels in this population.

## METHOD

### Subjects

Five hundred and seven potential subjects, between 25 and 60 years of age, were randomly chosen from over 20 000 patients on the lists of a group general practice in Abingdon (Oxfordshire). From this sample, 70 subjects were excluded who had moved or died; had severe psychosis, debilitating chronic illness, or chronic gastrointestinal disease; or were being treated for hyperlipidaemia or symptomatic coronary artery disease. The remaining 437 subjects were randomly assigned to either a control or a dietary intervention group and invited by telephone to participate in the study. Of these, 368 (84%) accepted the invitation and were enrolled.

## Study

All subjects completed a self-administered questionnaire concerning general health, smoking habits, and present diet. Those in the dietary intervention group were also given instruction regarding optimal body weight and diet by a nurse associated with the practice. This was done individually or in small groups, and took about 30 minutes per session. The dietary advice was directed towards a modest decrease in total fat intake from an expected level greater than 40% of calories to 30–35% of calories, with an increase in the ratio of polyunsaturated to saturated fats to approximately 0.4 from an expected level of less than 0.3. In addition, the value of increased dietary fibre, including soluble fibre,[4] was stressed. The potential benefits of physical exercise, and moderation of salt, alcohol, and tobacco intake were also mentioned, but not particularly emphasized. A booklet was given to intervention subjects which summarized the basic ideas of the diet, provided recipes, and offered advice concerning local restaurants. During a three month intervention period, the study nurse offered these subjects encouragement and advice regarding dietary modification. Promotional material was on display at the practice and brief follow-up/counselling sessions were scheduled for one and three months after entry to the study. The control group, on the other hand, were told that they were part of a nutrition survey, and were followed up on the same schedule by the same nurse, but without the dietary advice.

A fasting blood sample was obtained from each subject at entry, with repeated samples taken at one, three and 12 months after initial interview. Serum and plasma samples were processed promptly and frozen at $-20°C$ until analysis. Cholesterol concentrations were determined by an automated Liebermann–Burchardt reaction,[5] and lipoproteins were assayed by precipitation techniques.[6,7] Triglycerides were measured using a glycerokinase method,[8] and plasma glucose was determined by a glucose oxidase method (Boehringer). Triglycerides were not measured at one year. Triglyceride and cholesterol ester linoleic acid levels reflect dietary intake and thus were used to assess compliance with the diet. These were measured by methods as previously described[9] and expressed as the percentage of the total. Weight was measured by the study nurse.

A self-administered questionnaire developed by Gear and colleagues[10] was given at each encounter. This instrument used a simple food frequency format, and provided an accurate assessment of daily fibre intake. Although all aspects of diet (including alcohol) were reflected in the questionnaire, it was not designed to measure total calorie intake or to estimate precisely the intake of nutrients other than fibre. A separate brief questionnaire addressing attempts at dietary change was given to both groups at three months and one year. Intervention subjects were queried at one month, three months and one year about difficulties encountered with the recommendations.

## Analysis

Because of the known sex differences in lipid levels, all statistical analyses were done separately for men and women. Differences between means were evaluated for statistical significance by standard $t$-tests.[11] For baseline frequency data, statistical significance was determined by contingency table chi-square tests.[11]

## RESULTS

### Baseline characteristics

A total of 368 subjects [were] randomized into control (92 men, 89 women) or dietary intervention (97 men, 90 women) groups. In general, baseline characteristics were similar in the two groups (Table 1.1). However, control men had a higher percentage of current smokers than intervention men ($p < 0.05$), and a higher proportion of intervention women were in social class 1 or 2 compared with controls ($p < 0.01$). Intervention subjects tended to be heavier, though the differences were not

**Table 1.1** Baseline characteristics of subjects by sex and group assignment

| | Men | | Women | |
| --- | --- | --- | --- | --- |
| | Control group | Intervention group | Control group | Intervention group |
| No. of subjects | 92 | 97 | 89 | 90 |
| Mean age ± standard error (years) | 41.6 ± 1.0 | 42.1 ± 1.0 | 41.9 ± 1.1 | 41.1 ± 1.0 |
| Mean weight ± standard error (kg) | 76.3 ± 1.1 | 78.7 ± 1.2 | 62.6 ± 1.2 | 65.3 ± 1.4 |
| Mean height ± standard error (m) | 1.77 ± 0.01 | 1.77 ± 0.01 | 1.62 ± 0.01 | 1.63 ± 0.01 |
| % in social class 1 or 2 | 30.0 | 39.0 | 24.0 | 43.0** |
| % with diagnosis of hypertension | 14.0 | 12.0 | 15.0 | 8.0 |
| % who currently smoked | 48.0 | 32.0* | 30.0 | 28.0 |
| % who ever smoked | 74.0 | 67.0 | 57.0 | 49.0 |

* $p < 0.05$, ** $p < 0.01$ versus control group.

statistically significant. Women in the two study groups did not differ substantially with regard to history of gestational hypertension, hormone problems, hormone replacement therapy, use of oral contraceptives or parity.

In general subjects cooperated well with the study. Losses to follow up were modest, especially during the three month intervention period. Five subjects were unavailable at one month, 10 at three months (three control and seven intervention subjects) and 33 subjects (9.0%) were lost to follow up at 12 months (13 controls and 20 intervention subjects).

### Acceptance of diet

The dietary intervention appeared to be well accepted by the intervention group (Table 1.2). No one complained that the dietary advice was difficult to understand, and very few (at most 8%) thought the recommended regimen was hard to prepare or difficult to find in restaurants. However, approximately 10% of the intervention group noted that they or their families disliked the recommendations, and subjects with this complaint were more likely to drop out of the study.

### Reported changes in diet

At three months, more than two thirds of the diet group subjects reported consciously attempting to eat more fibre, compared with less than 2% of the controls (Table 1.3). There were similar large differences in the proportions attempting to reduce dietary fat, though reported efforts to increase intake of polyunsaturated fats were less marked. At one year, these trends continued, although there were some decreases in the proportion reporting continued efforts.

Reported dietary intake confirmed these patterns (Table 1.4). In contrast with controls, the intervention subjects reported dramatically increased intake of fibre and use of polyunsaturated fats, and decreased use of saturated fats. These patterns persisted at one year, though with some regression toward baseline values. There were no consistent differences between men and women with regard to uptake of the dietary recommendations.

The weights of the participants remained fairly stable at least

**Table 1.2** Difficulties encountered with the dietary advice among subjects randomized to the intervention group

|  | Men (No. of subjects*) % | | Women (No. of subjects*) % | |
|---|---|---|---|---|
| *Family or subject disliked diet* | | | | |
| 1 month | (95) | 11 | (87) | 9 |
| 3 months | (93) | 8 | (87) | 6 |
| *Diet hard to prepare* | | | | |
| 1 month | (95) | 2 | (87) | 1 |
| 3 months | (93) | 1 | (87) | 0 |
| 1 year | (83) | 0 | (83) | 4 |
| *Hard to eat out on diet* | | | | |
| 1 month | (95) | 4 | (87) | 2 |
| 3 months | (93) | 8 | (87) | 1 |
| 1 year | (83) | 5 | (83) | 5 |
| *Hard to understand diet* | | | | |
| 1 month | (95) | 0 | (87) | 0 |
| 3 months | (93) | 0 | (87) | 0 |
| 1 year | (83) | 0 | (83) | 0 |
| *Diet too expensive* | | | | |
| 1 month | (95) | 1 | (87) | 1 |
| 3 months | (93) | 0 | (87) | 1 |
| 1 year | (83) | 3 | (83) | 6 |

*Number of subjects evaluated vary because of losses to follow up and missing data.

during the first three months of participation, and at no time was there a significant difference between the two groups. There was a slight drop in the mean weight of the control group men after one year by 1.1 kg.

## Plasma lipid estimations

Changes in linoleic acid content of the circulating triglycerides and cholesterol esters were modest but consistent with the participants' reported increase in the dietary polyunsaturated: saturated fat ratio (Table 1.5).

Among the men, there were modest differences between diet groups in the changes in lipoproteins which generally

**Table 1.3** Reported efforts at dietary change by sex and group assignment

| | Men | | Women | |
|---|---|---|---|---|
| | Control (No. of subjects) % | Intervention (No. of subjects) % | Control (No. of subjects) % | Intervention (No. of subjects) % |
| *3 months* | | | | |
| Increased intake of fibre | (91) 1 | (93) 67 | (87) 2 | (86) 70 |
| Decreased intake of fat | (91) 1 | (93) 76 | (87) 1 | (86) 80 |
| Increased polyunsaturated fat | (91) 0 | (93) 29 | (87) 0 | (86) 53 |
| *1 year* | | | | |
| Increased intake of fibre | (86) 3 | (83) 52 | (79) 3 | (81) 42 |
| Decreased intake of fat | (86) 5 | (83) 55 | (79) 0 | (81) 38 |
| Increased polyunsaturated fat | (86) 1 | (83) 22 | (79) 1 | (81) 30 |

Differences between treatment groups were all statistically significant, $p < 0.001$.

paralleled the reported dietary changes (Table 1.5). By three months, total cholesterol declined slightly in men in the intervention group compared with a small increase in controls. Much of the reduction in the intervention group was due to a particularly large decrease in low density lipoprotein (LDL) cholesterol. In both groups, high density lipoprotein (HDL) cholesterol declined slightly during the three month diet period. By one year, the differences between the two groups of men had disappeared, with both showing reductions in total cholesterol and LDL cholesterol, and rises in HDL cholesterol. Among women there were no important differences between the diet groups at any time. HDL cholesterol tended to decrease in both groups. Analysis restricted to those in the highest quartile of total cholesterol (within sex group) was hampered by small numbers, but there were no statistically significant differences between treatment groups (data not shown).

## DISCUSSION

In this randomized controlled trial of dietary advice in a well British population, we found the dietary recommendations to be well accepted by a sample of adults registered in one group practice. Those randomized to receive the intervention found it understandable and affordable, and these subjects made few negative comments about the recommendations. High percentages of the intervention subjects reported increasing their intake of fibre and polyunsaturated fat and decreasing their intake of saturated fat. The long-term nature of the dietary change was particularly encouraging: despite the relatively brief intervention, there was substantial reported compliance with the recommendations at one year.

Triglyceride and cholesterol ester linoleic acid levels reflect dietary intake and thus were used to assess compliance with the diet. By these measures there was objective confirmation of the reported dietary patterns, although the changes in linoleic acid were small compared with those reported in another dietary intervention study (among subjects with hyperlipidaemia).[9] This smaller effect may be due to several factors, including our focus on a normal population, and the more moderate nature of our intervention.

Among men there were modest differences between the diet

**Table 1.4** Reported dietary consumption of fibre and fat by sex and group assignment

| | Men | | Women | |
|---|---|---|---|---|
| | Control | Intervention | Control | Intervention |
| *(No. of subjects) % using polyunsaturated fat for frying* | | | | |
| *(No. of subjects) mean total dietary fibre ± standard error (g per day)* | | | | |
| Baseline | (92) 19.3 ± 0.7 | (97) 20.4 ± 0.8 | (89) 16.4 ± 0.7 | (89) 18.9 ± 0.7 |
| 1 month | (92) 19.8 ± 0.8 | (95) 27.0 ± 1.0 | (88) 15.8 ± 0.6 | (88) 24.2 ± 1.0 |
| 3 months | (91) 21.1 ± 0.9 | (93) 27.8 ± 1.1 | (85) 15.7 ± 0.7 | (87) 24.8 ± 1.2 |
| 1 year | (69) 20.1 ± 1.0 | (56) 22.8 ± 1.0 | (68) 15.4 ± 0.8 | (65) 21.4 ± 1.0 |
| *(No. of subjects) % using polyunsaturated fat for spreading* | | | | |
| Baseline | (92) 12 | (97) 6 | (89) 11 | (90) 9 |
| 1 month | (92) 8* | (95) 67 | (88) 14* | (88) 74 |
| 3 months | (91) 8* | (93) 70 | (87) 12* | (87) 77 |
| 1 year | (87) 15* | (83) 58 | (81) 15* | (83) 54 |
| *(No. of subjects) % using polyunsaturated fat for frying* | | | | |
| Baseline | (92) 14 | (97) 14 | (89) 11 | (90) 10 |

|  | (No. of subjects) % | (No. of subjects) % | (No. of subjects) % | (No. of subjects) % |
|---|---|---|---|---|
| **using saturated fat for frying** | | | | |
| Baseline | (92) 23 | (97) 26 | (89) 19 | (90) 20 |
| 1 month | (92) 17* | (95) 4 | (88) 26* | (88) 7 |
| 3 months | (91) 19* | (93) 3 | (87) 25* | (87) 5 |
| 1 year | (85) 26* | (77) 9 | (79) 14* | (83) 7 |
| **using saturated fat for spreading** | | | | |
| Baseline | (92) 24* | (97) 41 | (89) 36 | (90) 31 |
| 1 month | (92) 28* | (95) 5 | (88) 34* | (88) 5 |
| 3 months | (91) 24* | (93) 3 | (87) 43* | (87) 0 |
| 1 year | (87) 23* | (83) 6 | (81) 37* | (83) 2 |

* $p < 0.05$ versus intervention group.

**Table 1.5** Fasting plasma lipids by sex and group assignment

| | Men | | Women | |
|---|---|---|---|---|
| | Control | Intervention | Control | Intervention |
| (No. of subjects) mean triglyceride linoleic acid ± SE (% of total) | | | | |
| Baseline | (91) 12.99 ± 0.43 | (92) 13.13 ± 0.50 | (85) 13.77 ± 0.66 | (86) 15.08 ± 0.63 |
| 1 month | (88) 12.29 ± 0.48* | (91) 15.39 ± 0.57 | (85) 12.98 ± 0.40* | (87) 15.13 ± 0.54 |
| 3 months | (89) 12.62 ± 0.51* | (90) 15.17 ± 0.61 | (85) 13.24 ± 0.43* | (85) 15.49 ± 0.55 |
| 1 year | (87) 12.65 ± 0.60* | (84) 14.52 ± 0.63 | (78) 13.94 ± 0.60* | (79) 15.98 ± 0.62 |
| (No. of subjects) mean cholesterol ester linoleic acid ± SE (% of total) | | | | |
| Baseline | (90) 42.59 ± 0.86 | (92) 40.80 ± 0.76 | (86) 44.70 ± 0.73 | (86) 43.56 ± 0.87 |
| 1 month | (88) 42.30 ± 0.90 | (91) 42.36 ± 0.90 | (86) 40.95 ± 0.70 | (87) 42.82 ± 0.79 |
| 3 months | (88) 38.58 ± 0.89 | (89) 40.51 ± 0.83 | (85) 41.87 ± 0.77 | (82) 43.22 ± 0.99 |
| 1 year | (84) 44.30 ± 0.82* | (81) 46.70 ± 0.79 | (78) 46.63 ± 0.95* | (77) 49.84 ± 0.83 |
| (No. of subjects) mean total cholesterol ± SE (mM) | | | | |
| Baseline | (92) 4.81 ± 0.08 | (97) 4.92 ± 0.08 | (89) 4.88 ± 0.10 | (89) 4.79 ± 0.09 |

*(No. of subjects) mean LDL cholesterol ± SE (mM)*

|  |  |  |  |  |
|---|---|---|---|---|
| Baseline | (80) 2.87 ± 0.09 | (85) 2.96 ± 0.08 | (87) 2.76 ± 0.10 | (84) 2.70 ± 0.09 |
| 1 month | (82) 2.77 ± 0.09 | (81) 2.73 ± 0.07 | (84) 2.81 ± 0.09 | (85) 2.77 ± 0.09 |
| 3 months | (81) 2.83 ± 0.08* | (89) 2.57 ± 0.08 | (81) 2.79 ± 0.11 | (77) 2.70 ± 0.09 |
| 1 year | (85) 2.31 ± 0.08 | (83) 2.36 ± 0.07 | (79) 2.73 ± 0.10 | (81) 2.71 ± 0.09 |

*(No. of subjects) mean HDL cholesterol ± SE (mM)*

|  |  |  |  |  |
|---|---|---|---|---|
| Baseline | (85) 1.36 ± 0.03 | (88) 1.33 ± 0.03 | (87) 1.67 ± 0.04 | (87) 1.64 ± 0.04 |
| 1 month | (86) 1.36 ± 0.03 | (84) 1.29 ± 0.03 | (86) 1.58 ± 0.04 | (86) 1.49 ± 0.03 |
| 3 months | (86) 1.32 ± 0.02 | (92) 1.29 ± 0.03 | (84) 1.51 ± 0.04 | (81) 1.44 ± 0.05 |
| 1 year | (86) 1.48 ± 0.03 | (84) 1.41 ± 0.03 | (79) 1.53 ± 0.03 | (81) 1.49 ± 0.03 |

*$p < 0.05$ versus intervention group. SE = standard error. LDL = low density lipoprotein. HDL = high density lipoprotein.

groups with regard to changes in lipoproteins. At the end of the three month diet period, the intervention group had experienced a significantly greater reduction in LDL cholesterol than the control group, though by one year the differences had narrowed. This suggests that the effects on lipoproteins may be strongest during the period of active encouragement of dietary change. Among women, there was little apparent impact of the diet programme, despite apparently similar levels of compliance among intervention subjects. Both intervention and control women experienced only minor changes in total and LDL cholesterol, with a slight fall in HDL cholesterol in both.

It is not clear why there was no apparent effect of the intervention among men at one year, or among women at any time, despite differences in reported diet similar to those among men in the first three months. The differences in linoleic acid content of cholesterol esters and circulating triglycerides suggest that this was not due simply to biased dietary reporting. One possible explanation for the results in men is the weight loss in the control subjects, which might have resulted in a lowering of LDL cholesterol in this group.[12] Also, it should be noted that other risk factor intervention studies have reported differences in the responses of men and women.[13-16] It is not clear what may underlie these differences, though hormonal factors are a possibility.

Several aspects of our study deserve comment. First, though the study population permits quite wide generalization, the results apply only to the particular intervention we employed. It is likely that different dietary advice or a different manner of motivating change might lead to different results. Secondly, the fact that the two groups were drawn from one geographical area and one practice may have permitted some of the control subjects to become aware of the intervention advice. This would lead to a conservative bias in our estimates of intervention effectiveness. Thirdly, our relatively small sample size provides only modest power for the detection of effects on lipids.

Finally, there were several differences in baseline characteristics between the two study groups, including lower baseline use of saturated fat for spreading among control men, higher social class among intervention women, and lower percentage of smokers among intervention men. Some of these differences may have been due to our relatively small sample size and the results of multiple comparisons. However, the differences could also have been due to selective recruitment. As noted above, subjects in the two treatment groups were given different explanations of the study at the first visit, and it is conceivable that the proportion cooperating thereafter varied differently in the two groups according to personal characteristics. For example, men who were smokers might have been willing to cooperate with the dietary survey presented to the control subjects, but not with the dietary change presented to the intervention group. This does not seem plausible, however, in light of the high (84%) acceptance rate among those invited to take part.

Previous investigations of dietary change in the primary prevention of coronary artery disease have employed various interventions. The earlier trials[3] used diets relatively high in fat (approximately 40% of calories) with polyunsaturated to saturated fat ratios greater than 1. More recently, interventions have been tested that employ diets somewhat lower in fat and with ratios of 0.4 to 0.8. (These have typically been in the setting of multifactorial trials.) In aggregate these have found that dietary change can be effective in lowering lipid levels, at least in high risk men. Our data show that current dietary recommendations made through general practice are acceptable to both sexes, but may have only limited efficacy, particularly among women. A larger, more detailed study will be required to document details of the effect.

## REFERENCES

1. National Advisory Committee on Nutrition Education. *Discussion paper on proposals for nutritional guidelines for health education in Britain*. London: Health Education Council, 1983.
2. Committee on Medical Aspects of Food Policy. Panel on diet in relation to cardiovascular disease. *Diet and cardiovascular disease*. London: Department of Health and Social Security, 1984.
3. Mann, J.I. and Marr, J.W. Coronary heart disease prevention: trials of diets to control hyperlipidaemia. In: Miller, N.E. and Lewis, B. (eds), *Lipoproteins atherosclerosis and coronary heart disease*, Chapter 12. Amsterdam: Elsevier, 1980.
4. Council on Scientific Affairs, Dietary fiber and health. *JAMA* 1989, **262**: 542–546.
5. Huang, T.C., Chem, C.P. and Wefler, V.A. Stable reagent for the Liebermann–Burchard reactions. Application for rapid serum cholesterol determinations. *Anal Chem* 1961, **33**: 1405–1407.

6. Burstein, M., Scholnick, H.R. and Morfin, R. Rapid method for the isolation of lipoproteins from human serum precipitation with polyions. *J Lipid Res* 1970, **11**: 583–595.
7. Ononogba, I.C. and Lewis, B. Lipoprotein fractionization by a precipitation method. A simple quantitative procedure. *Clin Chem Acta* 1976, **71**: 397–402.
8. Eggstein, M. and Kreutz, F.H. Eine neue Bestimming der Neutral-fette in Blutserum und Gewebe. *Klinische Wochenschrift* 1966, **44**: 262.
9. Moore, R.A., Oppert, S., Eaton, P. and Mann, J.I. Triglyceride fatty acids confirm a change in dietary fat. *Clin Endocrinol* 1977, **7**: 143–149.
10. Gear, L.S.S., Ware, A., Fursdon, P. *et al.* Symptomless diverti-cular disease and intake of dietary fibre. *Lancet* 1979, **1**: 511–514.
11. Armitage, P. *Statistical methods in medical research.* Oxford: Blackwell Scientific Publications, 1971.
12. Black, D., James, W.P.T., Besser, G.M. *et al.* Obesity: a report of the Royal College of Physicians. *J R Coll Physicians (Lond)* 1983, **17**: 5–65.
13. Puska, P., Salonen, J.T., Nissinen, A. *et al.* Change in risk factors for coronary heart disease during 10 years of a community inter-vention programme (North Karelia project). *Br Med J* 1983, **287**: 1840–1844.
14. Miettinen, M., Karvonen, M.J., Turpeinen, O. *et al.* Effect of cholesterol-lowering diet on mortality from coronary heart disease and other causes. *Lancet* 1972, **2**: 7782–7838.
15. Brownell, K.D. and Stunkard, A.J. Differential changes in plasma high-density lipoprotein-cholesterol levels in obese men and women during weight reduction. *Arch Intern Med* 1981, **141**: 1142–1146.
16. Hill, J.O., Thiel, J., Heller, P.A. *et al.* Differences in effects of aerobic exercise training on blood lipids in men and women. *Am J Cardiol* 1989, **63**: 254–256.

## 1.1 DISCUSSION

### 1.1.1 Objective

The objective of the trial was to measure the effects of dietary advice on the diets and fasting plasma lipids for a sample of men and women in a well population.

### 1.1.2 Design

An intervention group was given dietary advice directed towards a modest decrease in the total fat intake and an increase in the ratio of polyunsaturated to saturated fats. Written advice on diet and verbal encouragement were also given to the intervention group during a three-month period. A control group was not given dietary advice. The allocation of subjects to the intervention or control group was made randomly. All subjects completed a questionnaire on health, smoking habits and diet at the start of the study and all subjects were followed up after three months and also after one year.

### 1.1.3 Subjects

Of 368 subjects, chosen randomly from group practice lists in Abingdon, 97 men and 90 women were allocated to the intervention group, while 92 men and 89 women were allocated to the control group.

### 1.1.4 Outcome measures

- Baseline (initial) characteristics included age, weight, height, social class, hypertension and smoking habits.
- The percentage of the intervention group encountering one or more of five types of difficulty with dietary advice after one month, three months and one year.
- The percentages reporting efforts to change diets in specified ways after three months and one year.
- The reported daily consumption of fibre and fat at the beginning of the study, and after one month, three months and one year.
- The fasting plasma lipids at the beginning of the study, and after one month, three months and one year.

### 1.1.5 Data and statistical analysis

Statistical analysis was carried out on five sets of data (section 1.1.4) and Tables 1.1–1.5. For example, Table 1.1 quotes the means and standard errors for age, weight and height, while percentages are quoted for other baseline characteristics. Chapter 7 of this book discusses the definitions and appropriate use of 'summary statistics' such as mean, standard error and percentage.

The results of various hypothesis tests are noted in Tables 1.1–1.5; note the asterisks (*) relating to '$p$-values' below Tables

1.1, 1.3, 1.4 and 1.5. For example, $t$-tests have been carried out on the baseline numerical variables age, weight and height. In these the control group has been compared with the intervention group for men and also for women. Such 'unpaired' $t$-tests are discussed in Section 8.8, which includes an example from this case study. Further $t$-tests, 40 in all, were carried out in the case study using the data in Table 1.5.

Also from Table 1.1, percentages are compared, for example the control and intervention groups of men are compared for the percentages in social class 1 and 2. The test in these cases is the $\chi^2$ (chi-square) test, which is described in section 10.5 of this book, using an example from case study 2, but the idea is the same. Twelve chi-square tests were also used on the data in Table 1.3.

Which test(s) were used on the data from Table 1.4? This is left as an exercise for you to think about!

### 1.1.6 Further points

There are two points concerning the design and analysis of case study 1 which are worthy of comment. First – and this article is no different in this respect from the vast majority of articles in medical journals – no reason is given for the number of subjects included in the study. We are simply told that 507 potential subjects were selected and of these 368 actually took part in the trial. Why 507? Why not 50, or 5000 or 50 000? There are better ways of deciding 'study size' than choosing a number that is neither so small that everyone will agree that it is not large enough, nor so large that it takes too long or costs too much money to collect the data – see Chapter 14 for a discussion of 'study size' and 'sample size'.

Second, there are dangers in carrying out a large number of hypothesis tests. For example, a total of 14 hypothesis tests were carried out on the data in Table 1.1. It is not surprising that one of the tests showed a $p$-value of less than 0.05, since 0.05 implies 1 and 20, and implies that one test in 20 would give a $p$-value of less than 0.05 even if all the null hypotheses tested were, in fact, correct hypotheses. However, one of the tests, namely to compare the percentage of women in social class 1 or 2 for the intervention and control groups, resulted in a $p$-value of less than 0.01. This is small enough to make it very likely that the percentages in social class 1 or 2 really are different for the intervention and control groups. This makes social class a confounding factor with 'type of group' and makes any comparison of the control and intervention group fraught with danger in this case study.

Third, one might be tempted to question whether there is any point in comparing the baseline characteristics of the control and intervention groups in Table 1.3. Surely if the individuals in these groups were randomly allocated to one group or the other, then they are samples from the same populations (of weight, say). Hence the null hypotheses implied in Table 1.3 are all true and hence hypothesis tests are inappropriate. However, the above ignores the fact that, although 437 individuals were allocated randomly to one of the two groups, 16% of these in fact refused to enrol in the study. Hypothesis tests for baseline characteristics are therefore justified. On the other hand, this case study would have been better designed if the individuals who refused to enrol on the study had been identified *before* the allocation to 'control' or 'intervention'.

# 2

# Randomized controlled trial of anti-smoking advice by nurses in general practice

*A sick person intensely enjoys hearing of any material good,
any positive or practical success of the right. Do, instead of
advising him with advice he has heard at least 50 times before,
tell him of one benevolent act which has really succeeded
practically – it is like a day's health to him.*
From FN's publication Notes on Nursing *(1859–60)*

Case study 2 is an article by D. Sanders, G. Fowler, D.
Mant, A. Fuller, L. Jones and J. Marzillier taken from the
*Journal of the Royal College of General Practitioners*, **39** (1989),
273–6.

SUMMARY. Practice nurses are playing an increasingly promi-
nent role in preventive care, including the provision of anti-
smoking advice during routine health checks. A randomized
controlled trial was designed to assess the effectiveness of anti-
smoking advice provided by nurses in helping smokers to stop
smoking. A total of 14 830 patients aged 16–65 years from 11
general practices completed a brief questionnaire on general
health, incuding smoking status, at surgery attendance. The
doctor identified 4330 smokers and randomly allocated 4210 to
control or intervention groups. The doctor asked those in the
intervention group to make an appointment with the practice
nurse for a health check. The attendance rate at the health
check was 26%. Smokers were sent follow-up questionnaires at

one month and one year, and those who did not respond to
two reminders were assumed to have continued to smoke.
There was no significant difference in reported cessation between the intervention and control groups at one month or
one year. However, there was a significant difference in the
proportion of patients who reported giving up within one
month and who had not lapsed by one year: 0.9% in controls
and 3.6% in the intervention group ($p < 0.01$). Nevertheless,
the effect of the nurse intervention itself may be small as the
sustained cessation rate in attenders was only 42.4% higher
than in non-attenders. The deception rate in reporting cessation, as measured by urinary cotinine, was of the order of
25%.

## INTRODUCTION

Tobacco smoking is the most important cause of preventable
disease and premature death in developed countries[1] and control of cigarette smoking could achieve more than any other
single measure in the field of preventive medicine.[2] In the UK
smoking causes at least 100 000 premature deaths each year
and in 1984 the cost to the National Health Service of treating
smoking related diseases was estimated at more than £165
million.[3]

The great majority of those who smoke wish to stop and
many try to do so.[4] Mass media campaigns help to motivate
smokers to stop smoking, but are relatively ineffective in helping them to do so.[5] 'Smokers clinics' can offer effective help but
are few in number, attract only highly motivated smokers and
cannot, therefore, provide help on the scale required.[6] General
practitioners, on the other hand, see the majority of smokers
on their practice lists at least once a year and are expected by
their patients to take an active interest in behaviour that affects
health, including smoking.[7] Moreover, advice from general
practitioners has been shown to be effective in helping patients
to stop smoking[8-10] and adjuncts to verbal advice which may
enhance this effect include simple anti-smoking leaflets,[8]
demonstration of exhaled carbon monoxide[9] and nicotine
chewing gum, if properly used.[11] Consequently, primary health
care is widely acknowledged as being of vital importance
in health promotion generally,[12] and in smoking cessation in

particular.[13] However, in many practices it is no longer the doctor but the nurse who provides most preventive care, including asking and advising about smoking as part of health checks which are being widely promoted.[14] The effectiveness of anti-smoking advice given by nurses remains unproven and the nurses themselves have expressed a lack of confidence in the effectiveness of their role.[15]

This paper reports a randomized controlled trial designed to investigate the effectiveness of practice nurses in helping patients to stop smoking when invited to receive a brief health check.

## METHOD

The study took place in 11 general practices in the Oxford region, in which one or more of the nurses employed by the practice had expressed an interest in 'taking part in research on smoking' in a previous survey.[15] Before participating in the study each practice nurse received individual training in helping people to stop smoking, including attendance at two study days. List sizes varied from 3000 to 16500 and none of the practices had undertaken routine screening programmes of health checks previously. Only three of the practices undertook vocational training.

During the recruitment period, which varied in length according to the size of the practice, all 14830 patients aged 16 to 65 years attending surgery between Mondays and Fridays for an appointment with the doctor were asked to complete a questionnaire by the receptionist. This questionnaire included identifying details, demographic information, and brief questions on general health including smoking status. The patient gave the questionnaire to the doctor in the consultation. The 4330 smokers identified were intended to be allocated to a control or intervention group on a one to two basis according to the day of attendance. Although the doctors were given a desktop card to remind them which were control days and which intervention, 120 patients were allocated to the wrong group and were excluded from further analysis. The designation of specific days was itself randomized across weeks and practices, although the different recruitment rate in each practice meant that the exact 1:2 ratio was not achieved – 1310 controls and 2900 intervention patients were entered into the trial.

On control days, nothing further was done beyond usual care: the doctors were asked specifically not to discuss smoking beyond the requirement of the routine consultation. On intervention days, smokers were asked to make an appointment with the practice nurse for a health check, described as a routine check to assess blood pressure and weight and to discuss general health. Only 25.9% (751) of the 2900 patients in the intervention group made and kept an appointment with the practice nurse for a health check. A further 3.8% (109) made an appointment for a health check but did not attend. The number of patients who attended on designated intervention days and were not in fact asked to make an appointment by the general practitioner is unknown, but may account in part for the low attendance rate.

The 751 smokers who attended for the health check were further randomized to two equal sized groups: advice only (375 patients) and advice plus carbon monoxide test (376 patients).

During the heath check, blood pressure and weight were measured, family history of cardiovascular disease and cancer were discussed, and dietary and other health advice was given as necessary. The anti-smoking component consisted of advice and discussion, reinforced by written advice in the Health Education Council booklet *So you want to stop smoking?*, and the offer of a follow-up appointment. The same procedure was followed for patients allocated to the carbon monoxide group but in addition they were shown their level of expired air carbon monoxide using a Bedfont monitor, and its significance was discussed.

All attenders were followed up by a postal questionnaire at one month and one year. Random samples of one in two of the control group (642 patients) and of one in six of those who were randomized to the intervention group but did not attend for a health check (367 non-attenders) were similarly sent questionnaires one month and one year after their initial surgery attendance. Non-responders to the questionnaire were sent two reminders at intervals of three weeks.

In order to validate claimed smoking cessation, the patients in the control and attender groups who claimed to have stopped smoking at the one year follow up were invited for a further health check at which they were asked to provide a urine sample so that the level of cotinine, a metabolite of nicotine, in their urine could be measured. Four practices declined to participate.

Non-response to all three questionnaires at follow up was taken as an indication that the patient continued to smoke. Thus percentages of patients not smoking and confidence intervals were based on the number of patients in the group (rather than the number of questionnaire responders). The attender and non-attender groups were combined by weighting the non-attenders by the inverse of the sampling ratio. The $p$-values given are based on the $t$-test or chi-square test as appropriate. Confidence intervals are based on the standard error of a proportion.

## RESULTS

The mean age of the 751 attenders in the intervention group was 38.5 years while for the 2149 non-attenders it was 35.8 years ($p < 0.01$). There was also a significant difference in the proportion of attenders and non-attenders in social classes 1 or 2 (attenders 24.4%; non-attenders 29.9%, $p < 0.05$).

Of all 1760 smokers sent follow-up questionnaires only 59.2% completed them at both one month and one year; the response was similar in the controls (56.5%) and non-attenders (54.4%) but was significantly higher in the attenders (63.8%) ($p < 0.01$). The percentage of smokers in each study group who reported that they had stopped smoking when followed up is shown in Table 2.1. At neither one month nor one year follow up was there a significant difference in reported non-smoking between the intervention group and the controls. At one month there was a significant difference in reported non-smoking between the attenders and the non-attenders ($p < 0.05$), but not at one year. Surprisingly, the reported non-smoking rate was higher in all groups at one year than at one month.

The proportion of smokers who temporarily gave up smoking was far higher than those who achieved long term success (Table 2.1). In terms of the number of smokers reporting non-smoking at both one month and one year, and the number of smokers who claimed sustained cessation for one year, the intervention group performed significantly better than the

**Table 2.1** Self reports of non-smoking at follow up: comparison between intervention and control groups

| | Percentage not smoking* (95% confidence interval) | | | |
|---|---|---|---|---|
| | One month follow up | One year follow up | One month and one year follow up | Continuously from one month to one year follow up** |
| Controls (n = 642) | 5.3 (3.6–7.0) | 10.0 (7.7–12.3) | 1.2 (0.4–2.1) | 0.9 (0.2–1.7) |
| Intervention group | | | | |
| Attenders (n = 751) | 10.9 (8.7–13.1) | 12.9 (10.5–15.3) | 5.9 (4.2–7.6) | 4.7 (3.1–6.2) |
| Non-attenders (n = 367) | 6.5 (4.0–9.1) | 10.6 (7.5–13.7) | 4.1 (2.1–6.1) | 3.3 (1.4–5.1) |
| All (weighted average)† | 7.7 (5.7–9.7) | 11.2 (8.8–13.6) | 4.5 (3.0–6.0) | 3.6 (2.2–5.0) |

*Percentage of group total, assuming all non-responders still smoke. **Patients reporting non-smoking at one month and one year and who gave the date on which they last smoked as before the one month follow up. † Number of non-attenders weighted by inverse of sampling ratio (×5.9).

control group ($p < 0.01$). Moreover, the rate of sustained cessation in the non-attenders (3.3%) was intermediate to the rate in controls (0.9%) and attenders (4.7%) (chi-square trend 16.3, $p < 0.001$).

Table 2.2 shows the effect of adding carbon monoxide monitoring to the nurse health check. Although the percentage of patients who reported non-smoking at one month was slightly higher in the group receiving carbon monoxide monitoring, this difference was not statistically significant and the percentage reporting sustained non-smoking for one year was very similar (4.8% versus 4.5%).

Urine samples were obtained from 15 controls and 30 attenders who reported not smoking at the one year follow up. The cotinine assays indicated that eight of the attenders (26.7%, 95% confidence intervals 10.8–42.6%) and three of the controls (20.0%, 95% confidence intervals 0.0–40.2%) were regular or occasional smokers when they provided the urine sample. These deception rates are similar for patients who reported having given up at one year only (7/27, 25.9%) and for those who reported having given up at both one month and one year (4/18, 22.2%).

## DISCUSSION

An attempt to keep a formal record of whether a particular patient was asked to make an appointment by the general practitioners was abandoned early in the trial and, therefore, the extent to which the low attendance rate of 25.9% reflects a failure to offer a health check when appropriate is not known. Nevertheless, Pill and colleagues have recently reported a similarly low uptake of health checks by smokers with only 17% of attenders but 69% of non-attenders at health checks reporting that they had ever smoked.[16] If the majority of smokers are unlikely to attend health checks, then this in itself is an important limitation to the effectiveness of nurse anti-smoking advice at health checks. The observation that attenders were older than non-attenders suggests that this limitation may be particularly true for younger smokers.

The relatively high prevalence of not smoking at either one month or one year, but not both, underlines the need to measure outcome in terms of sustained cessation, as

**Table 2.2** Self reports of non-smoking at follow up: comparison between attenders according to use by nurse of a carbon monoxide meter

| | Percentage of attenders not smoking* (95% confidence interval) | | | |
|---|---|---|---|---|
| | One month follow up | One year follow up | One month and one year follow up | Continuously from one month to one year follow up** |
| Advice and CO monitoring (n = 376) | 11.7 (8.3–15.1) | 13.8 (10.1–17.5) | 5.9 (3.5–8.2) | 4.8 (1.6–8.0) |
| Advice only (n = 375) | 7.5 (4.4–10.5) | 14.7 (11.4–18.0) | 5.9 (3.5–8.2) | 4.5 (2.4–6.6) |

* Percentage of group total, assuming all non-responders still smoke. ** Patients reporting non-smoking at one month and one year and who gave the date on which they last smoked as before the one month follow up.

emphasized by the International Agency Against Cancer (UICC) guidelines on the conduct of trials of this nature.[13] There appears to be a population of smokers who frequently make transient attempts to stop smoking and render studies relying on single short-term outcomes difficult to interpret. In this study the difference in reported cessation between the control and intervention groups at one year (1.2%) is less than the difference in sustained cessation (2.7%) and this probably reflects random fluctuation in transient cessation.

In view of the low attendance rate for the health check it is of interest that there is a significant difference in sustained smoking cessation between the control and intervention groups. One explanation could be that the invitation by the general practitioner to make an appointment for a health check was itself an important anti-smoking intervention. This explanation is strengthened by the results of Russell and colleagues' study of minimal doctor intervention.[8] In Russell's 'questionnaire only' group, which is similar to our control group, the self-reported sustained cessation rate was 1.6%, while in his 'GP advice only' group, which is arguably similar to our non-attender group, the cessation rate was 3.3%. The fact that the cessation rate in our control group (0.9%) is lower than in Russell's study also raises the possibility that the general practitioners gave less advice to controls than they would normally, although this was contrary to the study protocol.

There is no doubt that self-reported cessation overestimates the true cessation rate – by about 25% in this study. Although the confidence intervals on the estimated deception rate are wide in this study the results are consistent with previous reports. The best validated recent study of self-reported smoking cessation, carried out by the British Thoracic Society, documented a deception rate of 27% at six months follow up and 25% at 12 months.[17] However, the self-reported cessation rate remains useful for comparative studies of effectiveness, as there is no evidence from this or previous studies that the deception rate is different in intervention and control groups.

The lack of effect of carbon monoxide monitoring is disappointing as a previous study had suggested that this might be helpful.[9] The relatively small numbers in the groups receiving advice with and without monitoring means that the power to exclude a small beneficial effect is limited, but there is no evidence for recent claims of a dramatic motivating effect.[18]

Trials of this type are fraught with methodological difficulty, and we have attempted to take a conservative approach throughout. However, the statistically significant difference between the intervention and control groups is dependent on the acceptance of a non-response to three questionnaires as a valid indication of continued smoking and the inclusion of the 25% of observations that may be deceptions. It should also be noted that the effect of the single nurse intervention described must be limited as the sustained cessation rate was only 42.4% higher in the attenders than in the non-attenders, despite the fact that the attenders are a selected compliant group.

Nevertheless, it must not be concluded that nurses cannot help smokers to stop smoking. It is quite possible that it is the context of the health check, at which a number of other measurements are made and at which other issues are discussed, which offers little scope for an effective nurse intervention. The cessation rate in the attender group was much lower than the smoking cessation rate recently reported by Richmond and Webster in Australia, which appears to have been achieved by intensive follow-up support of smokers as they gave up.[19] In view of our results, and the observation that about 10% of smokers claim to have temporarily stopped smoking at any one time, it is possible that the most appropriate role for the prevention nurse is not in giving initial advice to stop – which may be best done opportunistically by the general practitioner – but in the provision of longer term support and follow up which may be necessary to achieve sustained cessation.

## REFERENCES

1. World Health Organization. *Controlling the smoking epidemic.* WHO technical report series no 636. Geneva: WHO, 1979.
2. World Health Organization. *Smoking and its effect on health.* WHO technical report series no 568. Geneva: WHO, 1975.
3. Royal College of Physicians. *Health or smoking?* London: RCP, 1983.
4. Marsh, A. and Matheson, J. *Smoking attitudes and behaviour.* London: HMSO, 1983.
5. Dyer, N. *So you want to stop smoking: results at a follow-up one year*

*later*. London: British Broadcasting Corporation, 1983.

6. Chapman, S. Stop-smoking clinics: a case for their abandonment. *Lancet* 1985, **1**: 918–920.

7. Wallace, P.G. and Haines, A.P. General practitioners and health promotion: what patients think. *Br Med J* 1984, **289**: 534–536.

8. Russell, M.A.H., Wilson, C., Taylor, C. and Baker, C.D. Effect of general practitioners' advice against smoking. *Br Med J* 1979, **2**: 231–235.

9. Jamrozik, K., Vessey, M., Fowler, G. *et al.* Controlled trial of three different antismoking interventions in general practice. *Br Med J* 1984, **288**: 1499–1503.

10. Richmond, R.L., Austin, A. and Webster, I.W. Three year evaluation of a programme by general practitioners to help patients to stop smoking. *Br Med J* 1986, **292**: 803–806.

11. Lam, W., Sacks, H.S., Sze, P.C. and Chalmers, T.C. Meta-analysis of randomised controlled trials of nicotine chewing-gum. *Lancet* 1987, **2**: 27–30.

12. Secretaries of State for Social Services, Wales, Northern Ireland and Scotland. *Promoting better health* (Cm 249). London: HMSO, 1987.

13. Kunze, M. and Wood, M. (eds). *Guidelines on smoking cessation*. UICC technical report series. Volume 79. Geneva: UICC, 1984.

14. Fullard, E., Fowler, G. and Gray, M. Promoting prevention in primary care: controlled trial of low technology, low cost approach. *Br Med J* 1987, **294**: 1080–1082.

15. Sanders, D.J., Stone, V., Fowler, G. and Marzillier, J. Practice nurses and antismoking education. *Br Med J* 1986, **292**: 381–383.

16. Pill, R., French, J., Harding, K. and Stott, N. Invitation to attend a health check in a general practice setting: comparison of attenders and non-attenders. *J R Coll Gen Pract* 1988, **38**: 53–56.

17. British Thoracic Society. Comparison of four methods of smoking withdrawal in patients with smoking related diseases. *Br Med J* 1983, **286**: 595–597.

18. British Medical Association/Imperial Cancer Research Fund. *Help your patient stop*. London: BMA, 1988.

19. Richmond, R.L. and Webster, I.W. A smoking cessation programme for use in general practice. *Med J Aust* 1985, **142**: 190–194.

## 2.1 DISCUSSION

### 2.1.1 Objective

The trial was designed to assess the effectiveness of anti-smoking advice provided by nurses to help smokers to stop smoking.

## 2.1.2 Design

Smokers attending surgery assigned to an intervention group were asked by their general practitioner to make an appointment with the practice nurse for a health check. Those who attended were equally divided at random into two sub-groups (advice only and advice with a carbon monoxide (CO) test). Smokers attending surgery who formed the control group were given no special anti-smoking advice by their GP. The allocation of subjects to the control or intervention group was done systematically by designating some surgery days as 'control' and some as 'intervention' during the recruitment period with the aim of achieving a 1:2 ratio. Postal questionnaires about smoking habits were sent after one month and after one year to the following:

- all attenders for the health check in the intervention group;
- a proportion (1 in 6) of non-attenders in the intervention group;
- a proportion (1 in 2) of the control group.

Urine samples were taken from some subjects to validate claimed smoking cessation.

## 2.1.3 Subjects

Of 2900 allocated to the intervention group, 751 (25.9%) attended for a health check while 2149 did not attend. Of the attenders, 375 were given advice only while 376 were also given a CO test. The control group consisted of 1310 subjects. Those followed up by a postal questionnaire after one month and one year comprised a total of 1760 smokers as follows:

- 751 intervention group attenders;
- 367 intervention group non-attenders;
- 642 of the control group.

## 2.1.4 Outcome measures

- Baseline characteristics were age and social class.
- The main outcome measure was the 'percentage not smoking' in various sub-groups of subjects, at one month, at one year, and at both one month and one year, and also

whether the cessation was sustained continuously from one month to one year follow-up.

- Urinary cotinine to help decide whether those reporting not smoking were actually non-smokers.

### 2.1.5 Data and statistical analysis

Although not reported in either of the tables in this case study, the mean ages of the 751 attenders and the 2149 non-attenders were compared in its *Results* section using a *t*-test (section 8.8). The proportion (and hence the percentages) of those in social class 1 and 2 for attenders and non-attenders were also compared by means of a chi-square test (section 10.5).

Turning now to the data in Tables 2.1 and 2.2, we find percentages quoted with their associated 95% confidence intervals; section 10.2 contains a numerical example drawn from Table 1.1 (i.e. controls, one month follow-up). The percentage of attenders and non-attenders not smoking at the one-month follow-up can be compared in two ways, namely using the 'confidence interval approach' of section 10.4, or by using the 'hypothesis-test approach' which in this case is the chi-square test of section 10.5. Both approaches have been used on the relevant data from Table 2.1, and are detailed in sections 10.4 and 10.5, respectively. Other comparisons between pairs of percentages are reported in the *results* section of case study 2. The last comparison which refers to Table 2.1 quotes the chi-square trend test. This test is described in section 11.2 and the example given there refers to the relevant data from Table 2.1. The data from Table 2.2 have also been analysed using standard chi-square tests, but no significant differences were found.

### 2.1.6 Further points

The authors of this case study admit that there are difficulties in drawing useful conclusions from the study for four reasons:

- For 'unknown reasons' only 25.9% allocated to the intervention group actually attended for a health check.
- 'There seems to be a proportion of smokers who make frequent transient attempts to stop smoking'. For example, the percentage quoting sustained cessation from one month

to one year is only 3.6% for the intervention group, but it is up to 11.2% for those who were not smoking at the one-year follow-up for the same group.

- The deception rate is of the order of 25% for those reporting not smoking.
- The assumption made about those who did not respond to three questionnaires, namely that they continued to smoke.

In spite of these difficulties the authors conclude, in a double-negative statement, that 'it must not be concluded that nurses cannot help smokers to stop smoking'. A more objective statement might be appropriate: for example, 'a survey of this type is unlikely to demonstrate whether (or not) nurses can help smokers to stop smoking'.

# 3

# Psychological distress: outcome and consultation rates in one general practice

Case study 3 is an article by Alastair F. Wright taken from the *Journal of the Royal College of General Practitioners*, **38** (1988), 542–5.

SUMMARY. *This paper reports a one-year follow-up of random samples of 90 male and 96 female patients attending one general practitioner. There was no statistically significant difference between men and women in the total score on the 28-item general health questionnaire or any of the subscores. However, the diagnostic labels applied to the two sexes were strikingly different as was the prescribing of psychotropic drugs. Outcome of psychological distress was assessed in terms of change in total general health questionnaire score. Two thirds of the patients (65%) showed normal scores at the beginning and end of the follow-up period, 19% changed from abnormal to normal and 8% changed from normal to abnormal. The remaining 9% had persistently high scores though less than half had been given a psychiatric diagnosis. They had very high consultation rates persisting over several years and three-quarters were known to have chronic physical illness. It seems possible that some patients with persistently high consultation rates who present with chronic, mainly somatic, symptoms may be or may become psychologically distressed to a significant degree and that this psychological distress goes unrecognized in the presence of physical disease.*

## INTRODUCTION

Relatively few studies of the outcome of psychiatric illness have been carried out in the primary care setting,[1] partly because of problems of subjectivity in such research but mainly because psychiatric illness is difficult to classify, particularly in the community.

Doctors have different perceptions of what constitutes psychiatric illness but problems of observer bias can be partly overcome by the use of self-report measures of psychiatric symptomatology such as the general health questionnaire.[2-6] The use of a questionniare is often more acceptable to the patient than to the doctor[7] and can unmask psychiatric illness in patients who avoid presenting emotional symptoms to their general practitioner.[8-11]

The general health questionnaire is a self-reporting screening questionnaire which identifies individuals who have a high probability of suffering from psychiatric illness.[12] It has high reliability and correlates well with the clinical assessments of consultant psychiatrists.

In a study of emotional disturbance in newly-registered general practice patients, Corser and Philip[13] found that high-scorers on the general health questionnaire had more episodes of illness, more severe ratings of psychiatric problems and were more likely to have a formal psychiatric diagnosis. Goldberg and Bridges[14] pointed out that the higher the general health questionnaire score the more likely it was that the patient could be diagnosed and the less likely it was for the disorder to remit spontaneously. Johnstone and Goldberg[15] demonstrated the efficacy of the general health questionnaire in the secondary prevention of minor psychiatric morbidity in general practice and found the effects of detection to be 'beneficial and immediate'.

Patients in general practice often present problems which are a mixture of physical, psychological and social elements and these problems may be transient or represent illness in its earliest stages. This and the continuing relationship with patients and their families may lead the general practitioner to a different perspective of illness and a different therapeutic decision from his hospital colleague.[16] The continuing relationship may also serve to emphasize awareness of the suffering which patients experience in the psychological and social aspects of their lives and justify general practice management of psychological distress as opposed to diagnosed mental illness.

The general health questionnaire is an acceptable here-and-now measure of emotional distress[13] but a raised score does not necessarily equate with a clinical diagnosis of psychiatric illness. Also, distress is not felt only by the mentally ill and the management of this distress is not the responsibility of the psychiatrist alone.

This paper reports a one-year follow-up by a single observer of random samples of male and female patients attending one general practitioner. Outcome of psychological distress was assessed in terms of the change in total general health questionnaire score with analysis of consultation rates and prescribing of psychotropic drugs. The aim of the study was to determine whether patients with persistently distressing psychological symptoms (as measured by the total score on the general health questionnaire) constitute a distinguishable group with higher than average consultation rates.

## METHOD

Using a table of random numbers, samples were drawn from patients aged over 17 years and under 65 years attending one general practitioner in a group practice over a period of five months. The results of the original study involving these patients have already been reported.[17] Samples were selected so that all the consulting sessions in a week were represented by a 10% random sample of patients attending that session and no patient was included more than once. Home visits, which account for approximately 8% of the workload, were excluded.

Patients were asked to complete the 28-item general health questionnaire while waiting to see the doctor and to answer questions on social and employment status. Social status was determined using the list in the Royal College of General Practitioner's *Classification and analysis of general practice data.*[18]

A record was kept of the total number of consultations, referrals and psychotropic drugs prescribed, by any of the six doctors in the practice, for one year after recruitment at which time the patient was requested to complete a second general health questionnaire without seeing the doctor. In order to identify which patients had higher than average consultation rates the total number of consultations in the five years prior to recruitment was also determined for each patient by a retrospective search of the records.

### Statistical tests

For data recording the numbers of patients with a given attribute the chi-square test was used to test for the significance of the difference between two proportions, using Yates's correction for two by two tables. When each patient had a score on a variable, for example general health questionnaire score, the *t*-test or one-way analysis of variance was employed.

Using the sampling method described the probability of an individual being selected is proportional to the indvidual's consulting rate. Therefore, statistical tests were repeated where appropriate after reweighting to compensate for this. This recalculation, together with a study of scatter diagrams, did not suggest that the results had been biased by the sampling scheme. In the interests of simplicity, only the unweighted results are presented here.

### RESULTS

Ten of the 115 men in the original sample[17] had left the practice during the follow-up year. Of the remaining 105 men, 14 did not respond and one returned an incomplete second questionnaire, giving a valid response from 90 men (86% response rate). Similarly, five of the 112 women in the original sample had left. Of the remaining 107, 10 did not respond and one returned an incomplete questionnaire, giving a valid response from 96 women (90% response rate). There were no statistically significant differences between responders and non-responders in respect of age, social status, total number of consultations or number of consultations with a psychiatric diagnosis during the follow-up year.

### Age and social status of respondents

The mean age of the men was 43.3 years (range 19–64 years) and of the women 37.1 years (range 18–64 years). The percenage distribution of the respondents by social status was professional 2%, intermediate 6%, skilled non-manual 23%, skilled manual 24%, semi-skilled 30% and unskilled 15%.

## Mean initial scores

There were no statistically significant differences between the men and women in total general health questionnaire score or any of the subscores at the beginning of the study (Table 3.1).

## Consultation rates

The 90 men had a total of 597 consultations over the one-year period and the 96 women had 850, of which 108 were for antenatal care only. The mean total consultation rate of 8.9 (standard error 0.7) for the women was significantly higher than that for the men (6.6, SE 0.5, $t$-test, $p < 0.02$), but when antenatal consultations were excluded the rate for the women became 7.7 (SE 0.6) and the difference was no longer statistically significant. Fifty-seven men (63%) and 47 women (49%) were known to be chronic/recurrent health problems (chi-square = 3.9, 1 degree of freedom, $p < 0.05$). Twenty men (22%) received a diagnosis of psychiatric illness at least once in the follow-up year compared with 35 women (36%) (chi-square = 4.5, 1 df, $p < 0.05$). There were 141 psychiatric consultations for the 20 men in the follow-up year (mean 7.1) and 178 for the 35 women (mean 5.0) but this difference was not significant.

**Table 3.1** Mean scores on the general health questionnaire (GHQ) at the beginning of the study

|  | *Mean GHQ score (standard error)* | |
|---|---|---|
|  | *Men (n = 90)* | *Women (n = 96)* |
| Total | 5.7 (0.7) | 6.8 (0.7) |
| Somatic symptoms | 6.4 (0.5) | 7.0 (0.5) |
| Anxiety | 6.1 (0.5) | 6.4 (0.5) |
| Social disturbance | 8.5 (0.4) | 8.9 (0.3) |
| Severe depression | 2.9 (0.5) | 2.3 (0.4) |

$n$ = number of respondents.

### Diagnoses and prescribing

The diagnostic labels of psychiatric illness applied to the two sexes were significantly different (Table 3.2) and there were similar differences in the numbers of patients of each sex who received psychotropic drugs at least once in the follow-up year (Table 3.3).

### Outcome

Outcome was assessed by comparing the general health questionnaire scores at the beginning and at the end of the follow-up year (Table 3.4). There was no significant difference in the outcome pattern between men and women. Two-thirds of patients (65%) showed normal scores at the beginning and the end of the follow-up period and 9% had persistently high scores. The patients were divided into four groups –

**Table 3.2** Diagnostic labels given to the patients receiving a diagnosis of psychiatric illness in the follow-up year

|  | Number (%) of patients | |
|---|---|---|
|  | Men (n = 20) | Women (n = 35) |
| Depression | 13 (65) | 10 (29) |
| Anxiety state | 2 (10) | 18 (51) |
| Other | 5 (25) | 7 (20) |

$\chi^2 = 10.2$, 2 df, $p < 0.01$. $n$ = total number of patients.

**Table 3.3** Psychotropic drugs prescribed to the patients receiving a diagnosis of psychiatric illness in the follow-up year

|  | Number (%) of patients | |
|---|---|---|
|  | Men (n = 20) | Women (n = 35) |
| Antidepressants | 9 (45) | 12 (34) |
| Benzodiazepine anxiolytics* | 4 (20) | 17 (49) |
| Both | 4 (20) | 6 (17) |
| No drugs prescribed | 3 (15) | 0 (0) |

$\chi^2 = 8.4$, 3 df, $p < 0.05$. $n$ = total number of patients.
*Benzodiazepines are often prescribed for relatively short periods and these figures do not necessarily reflect long-term usage.

**Table 3.4** Comparison of total general health questionnaire scores at the beginning and at the end of the follow-up year

| Total GHQ score at beginning | Total GHQ score at end | Mean number (%) of patients Men (n = 90) | Women (n = 96) |
|---|---|---|---|
| ≤8 (normal) | ≤8 (normal) | 61 (68) | 59 (61) |
| >8 (abnormal) | >8 (abnormal) | 7 (8) | 10 (10) |
| >8 (abnormal) | ≤8 (normal) | 15 (17) | 20 (21) |
| ≤8 (normal) | >8 (abnormal) | 7 (8) | 7 (7) |

n = total number of patients.

normal/normal, abnormal/abnormal, abnormal/normal and normal/abnormal. Normal represents a total general health questionnaire score of eight or less while abnormal represents a score of more than eight. A cut-off point of nine was chosen as previous work in the same practice[17] had indicated that this threshold gave the best trade-off between sensitivity and specificity in this practice population. Plots of the data confirmed that this choice of cut-off score had not significantly biased the results. The four groups identified were also compared in terms of the number of patients given a psychiatric diagnosis during the year, prescribed a psychotropic drug during the year and known to suffer a chronic illness (Table 3.5). In the group with persistently abnormal scores less than half had been given a psychiatric diagnosis or prescribed psychotropic drugs. This group had the highest percentage of patients with known chronic physical illness.

The consultation rates for the study year and the five years before recruitment were determined for each of the four groups defined above (Table 3.6). Patients with persistently abnormal general health questionnaire scores also showed high consultation rates persisting over several years.

## DISCUSSION

In using the sampling method described it was recognized that the probability of an individual being selected would be proportional to the individual's consulting rate. While this problem could have been avoided by random sampling from

**Table 3.5** Comparison of the groups by the number given a psychiatric diagnosis during the year, prescribed a psychotropic drug during the year and known to suffer a chronic illness

| | Number (%) of patients | | | |
| | Normal/ normal (n = 120) | Abnormal/ abnormal (n = 17) | Abnormal/ normal (n = 35) | Normal/ abnormal (n = 14) |
| --- | --- | --- | --- | --- |
| Given psychiatric diagnosis | 13 (11) | 7 (41) | 19 (54) | 6 (43)* |
| Given psychotropic drug | 15 (12) | 7 (41) | 22 (63) | 6 (43)** |
| Known to suffer chronic non-psychiatric illness | 55 (46) | 13 (76) | 20 (57) | 4 (29)† |
| Known to suffer chronic psychiatric illness | 6 (5) | 2 (12) | 4 (11) | 3 (21) |

$n$ = total number of patients. * $\chi^2$ = 34.3, 3 df, $p < 0.001$. ** $\chi^2$ = 39.3, 3 df, $p < 0.001$. † $\chi^2$ = 8.9, 3 df, $p < 0.05$.

the practice list, this solution was rejected as the lists of individual doctors are not kept separate and patients are free to consult any of the partners or the trainee as they wish. In addition, the population of interest was 'attending' patients and sampling from the list of patients at risk would have resulted in a lower response rate.

No statistically significant differences were found between men and women in mean initial scores on the general health questionnaire or in any of the subscales, yet the diagnostic labels applied to those thought to be suffering from psychiatric illness were very different and women were more than twice as likely to receive benzodiazepine anxiolytics as men. The prevalence of diagnosed psychiatric illness was significantly higher in women than men, which is in keeping with most published data; Briscoe[19] suggests that women are more likely to express their feelings, both pleasant and unpleasant, than men.

On the other hand the men with psychiatric illness had on average more consultations with a psychiatric diagnosis than the ill women. Men were also more likely than women to have chronic or recurrent health problems (63% versus 49%). Though a small series, there is some support for the clinical speculation that the ill men seen were on average more dis-

Table 3.6 Mean consultation rates in the year of the study and in the five preceding years for the four groups of patients

| | Five years before study | Four years before study | Three years before study | Two years before study | Year before study | Study year |
|---|---|---|---|---|---|---|
| | Mean number of consultations per patient per year (number of patients*) | | | | | |
| Women** | | | | | | |
| Normal/normal | 5.5 (51) | 4.5 (52) | 4.9 (53) | 5.6 (54) | 6.9 (57) | 6.2 (59) |
| Abnormal/abnormal | 8.3 (8) | 8.5 (8) | 7.9 (8) | 11.1 (8) | 13.0 (9) | 13.2 (10) |
| Abnormal/normal | 5.6 (19) | 6.5 (19) | 5.8 (19) | 6.3 (19) | 8.6 (20) | 9.3 (20) |
| Normal/abnormal | 7.8 (6) | 5.1 (7) | 7.3 (7) | 6.0 (7) | 5.6 (7) | 8.0 (7) |
| | NS† | NS | NS | NS | $p < 0.05$ | $p < 0.01$ |
| Men | | | | | | |
| Normal/normal | 2.9 (55) | 3.3 (55) | 3.0 (56) | 4.1 (57) | 4.4 (58) | 5.5 (61) |
| Abnormal/abnormal | 5.0 (6) | 6.3 (6) | 2.4 (6) | 7.3 (7) | 9.0 (7) | 11.4 (7) |
| Abnormal/normal | 3.6 (12) | 5.5 (13) | 8.1 (15) | 8.1 (15) | 7.7 (15) | 9.1 (15) |
| Normal/abnormal | 2.4 (7) | 2.7 (7) | 3.3 (7) | 3.4 (7) | 5.7 (7) | 6.1 (7) |
| | NS | NS | $p < 0.01$ | $p < 0.05$ | NS | $p < 0.05$ |

*Number of patients who were registered with the practice for the whole of the year and so had complete record of attendance.
**Antenatal consultations excluded. † Analysis of variance.

turbed and needed to be seen more frequently than the ill women. It may be, however, that men in the practice are less likely to consult for psychological symptoms than women and the sample of consulting men may be biased by the inclusion of a high proportion of men with chronic health problems, representative of consulting men, but not of men within the practice.

Dividing the patients into four outcome groups according to the change in their general health questionnaire score supported the hypothesis that patients with persisting high levels of distressing psychiatric symptoms (abnormal/abnormal) constitute a distinct group with higher consultation rates than the other groups. These high rates were found to persist over several years. The group which may represent patients recovering from mental disturbance (abnormal/normal), and the group which may represent patients becoming mentally disturbed (normal/abnormal) both showed a more moderate and less persistent rise in consultation rate. The slight increase in consulting frequency in the years prior to recruitment seen in the patients who could be regarded as ill/becoming well (abnormal/normal) is in keeping with the observations of Widmer and Cadoret in the USA,[20] who commented that a 'pattern of increased office visits with a constellation of varied functional somatic complaints' often indicates that depression is developing.

The high demands of the group with persisting psychiatric symptoms have important implications for workload and patient care and Buchan and Richardson[21] have shown that consultations with such patients take longer than average, especially when they are follow-up visits. Less than half of this group were given a psychiatric diagnosis during the follow-up year but the percentage of patients with chronic non-psychiatric illness was high.

It seems possible that some patients with persistently high consultation rates who present with chronic, mainly somatic, symptoms may be or may become psychologically distressed to a significant degree and that this psychological distress goes unrecognized in the presence of physical disease. Knox and Neville in a survey in a similar practice showed that 14% of consultations with a non-psychiatric diagnosis had significant psychiatric content and that there was likely to be some under-reporting owing to psychiatric disturbance going unrecognized (unpublished results). Goldberg and Bridges,[14] studying new episodes of psychiatric illness in the community, pointed out that most were associated with physical illness or were somatized presentations of psychiatric illness. They found pure psychiatric presentations to be quite rare, accounting for only 5% of new illness. A high index of suspicion is required to detect psychiatric illness in patients with known physical illness and here the general health questionnaire, which is simple to use and acceptable to patients, would appear to have a clinically useful role as a probability estimate of caseness.[22]

While the findings of this study do not constitute proof, they justify more detailed investigation of consulting patterns in patients with psychiatric illness, with symptoms such as giddiness, lassitude or multiple aches and pains which may be psychosomatic and also with major chronic or recurring health problems. In addition to clinical assessment and questionnaire measures of psychiatric symptomatology, investigations should include a standard measure of patient personality, such as the Eysenck personality questionnaire,[23] and an assessment of factors which are perceived by the patient to be social problems.

## REFERENCES

1. Wilkinson, G. *Overview of mental health services in primary care settings, with recommendations for further research.* Washington: US Department of Health and Human Services, 1986.
2. Goldberg, D.P. *The detection of psychiatric illness by questionnaire.* London: Oxford University Press, 1972.
3. Goldberg, D.P. Identifying psychiatric illness among general medical patients. *Br Med J* 1985, **291**: 161–162.
4. Sims, A.C.P. and Salmons, P.H. Severity of symptoms of psychiatric outpatients: use of the general health questionnaire in hospital and general practice patients. *Psychol Med* 1975, **5**: 62–66.
5. Overton, G.W. and Wise, T.N. Psychiatric diagnosis in family practice: is the general health questionnaire an effective screening instrument? *South Med J* 1980, **73**: 763–764.
6. Tarnopolsky, A., Hand, D.J., McLean, E.K. *et al.* Validity and uses of a screening questionnaire (GHQ) in the community. *Br J Psychiatry* 1979, **134**: 508–515.
7. Short, D. Why don't we use questionnaires in the medical outpatient clinic? *Health Bull (Edinb)* 1986, **44**: 228–233.
8. Skuse, D. and Williams, P. Screening for psychiatric disorder in general practice. *Psychol Med* 1984, **14**: 365–377.

9. Goldberg, D.P. and Blackwell, B. Psychiatric illness in general practice. A detailed study using a new method of case identification. *Br Med J* 1970, **2**: 439–443.
10. Marks, J., Goldberg, D.P. and Hillier, V.E. Determinants of the ability of general practitioners to detect psychiatric illness. *Psychol Med* 1979, **9**: 337–353.
11. Goldberg, D.P., Steele, J.J., Johnson, A. *et al.* Ability of primary care physicians to make accurate ratings of psychiatric symptoms. *Arch Gen Psychiatry* 1982, **39**: 829–833.
12. Goldberg, D.P. *Manual of the general health questionnaire.* Windsor: NFER, 1978.
13. Corser, C.M. and Philip, A.E. Emotional disturbance in newly registered general practice patients. *Br J Psychiatry* 1978, **132**: 172–176.
14. Goldberg, D. and Bridges, K. Screening for psychiatric illness in general practice: the general practitioner versus the screening questionnaire. *J R Coll Gen Pract* 1987, **37**: 15–18.
15. Johnstone, A. and Goldberg, D. Psychiatric screening in general practice. A controlled trial. *Lancet* 1976, **1**: 605–608.
16. Howie, J.R.G. Diagnosis – the Achilles heel? *J R Coll Gen Pract* 1972, **22**: 310–315.
17. Wright, A.F. and Perini, A.F. Hidden psychiatric illness: use of the general health questionnaire in general practice. *J R Coll Gen Pract* 1987, **37**: 164–167.
18. Royal College of General Practitioners. *Classification and analysis of general practice data.* Occasional paper 26. London: RCGP, 1986.
19. Briscoe, M. Sex differences in psychological well-being. *Psychol Med* 1982 (Monogr Suppl 1).
20. Widmer, R.B. and Cadoret, R.J. Depression in primary care: changes in pattern of patient visits and complaints during developing depressions. *J Fam Pract* 1978, **7**: 293–302.
21. Buchan, I.C. and Richardson, I.M. *Time study of consultations in general practice.* Scottish Health Services studies no. 27. Edinburgh: Scottish Home and Health Department, 1973.
22. Goldberg, D. Use of the general health questionnaire in clinical work. *Br Med J* 1986, **293**: 1188–1189.
23. Eysenck, H.J. and Eysenck, S.B.G. *Manual of the Eysenck personality questionnaire.* Sevenoaks: Hodder & Stoughton, 1984.

## 3.1 DISCUSSION

### 3.1.1 Objective

The aim of the study was to determine whether patients with persistently distressing psychological symptoms (as measured by the total score on the general health questionnaire (GHQ)) constitute a distinguishable group with higher than average consultation rates.

### 3.1.2 Design

A sample of patients attending their GP in a group practice completed the 28-item GHQ and answered questions on social and employment status. Also recorded were consultations, referrals and psychotropic drugs prescribed for the one-year period after recruitment. After a year a second GHQ was completed. Historical five-year consultation rates were also determined.

### 3.1.3 Subjects

Of an original sample, a few dropped out or did not complete the questionnaire satisfactorily, leaving 90 mean and 96 women. Of these, 20 men and 35 women received a diagnosis of psychiatric illness during the follow-up year.

### 3.1.4 Outcome measures

- Baseline characteristics were sex, age, social class, as well as the total score and sub-scores for the GHQ at the beginning of the follow-up year.
- Consulation rates in the follow-up year were compared for all 90 men and 96 women, and for the 20 men and 35 women referred to above.
- Diagnostic labels and types of psychotropic drug given to the 20 men and 35 women.
- The GHQ score at the end of the follow-up year.

### 3.1.5 Data and statistical analysis

In Table 3.1, means and standard errors are reported, separately for the 90 men and 96 women, for the total of GHQ score and four sub-scores. Of the five comparisons between male and female data we are told none was significant. This conclusion could be checked using unpaired $t$-tests (section 8.8). Results of two more $t$-tests are reported in the subsection on 'Consultation rates'.

In the same sub-section, two chi-square tests (section 10.5) are reported, but checking indicates that Yates's correction was not included in either test. Had it been, the conclusion of one of the tests would have been different!

For the data from Table 3.2, which is a 3 × 2 contingency table, another chi-square test was used correctly, but the same test should not have been applied to the '4 × 2' Table 3.3, since three out of eight expected values are below 5 (section 10.5). The correct test here is the Fisher exact test (section 11.1). The correct conclusion for the data in Table 3.3 is stated in section 11.1.

For the data in Table 3.4, men and women are again compared, presumably using a chi-square test, although neither a statistic nor a $p$-value is quoted.

From *each* row of Table 3.5 it is possible to form a 2 × 4 contingency table, for example the first table would have rows labelled 'given psychiatric diagnosis' (observed 13, 7, 19, 6), and 'not given psychiatric diagnosis' (observed 107, 10, 16, 8). However, for the bottom row the expected frequencies are too small to analyse (left for the reader to confirm!).

In Table 3.6, four groups of female patients are compared at six points in time, along with four male groups at the same points in time. Each analysis has been performed using a technique called analysis of variance (ANOVA) described in Chapter 9. It is not possible to check the conclusions reached in Table 3.6 since insufficient information is given (one would need the actual number of consultations of each of the 96 women and 90 men in each time period). However, a similar example is described in section 9.3.

### 3.1.6 Further points

There seems to be a general overstating of results in this study. For example, in the section on consultation rates, only one test in four gave a $p$-value below 0.05, once antenatal consultations were excluded and Yates's correction had been used correctly.

A significant effect was correctly found in Table 3.2. For the data in Table 3.3, the wrong test was used, but fortunately the correct conclusion (of a significant effect) was drawn in spite of this mistake.

Table 3.6 is used to draw the conclusion that 'patients with persistently high levels of distressing psychiatric symptoms (abnormal/abnormal) constitute a distinct group with higher consultation rates than the other [three] groups'. While it is true that the null hypothesis that 'the mean consultation rates for the four groups are equal' was rejected in five cases out of 12, this does not necessarily mean that the group with the arithmetically highest mean is significantly higher than the other three group means. One needs to do a posterior test to refine the conclusions of ANOVA (a similar example is shown in section 9.3). No posterior tests were reported for the data in Table 3.6. However, inspection of the means in the five significant cases shows that the mean for the abnormal/abnormal group was arithmetically the highest in only three of the five cases. Not so convincing as it appears!

# 4

# Use of regression analysis to explain the variation in prescribing rates and costs between family practitioner committees

Case study 4 is an article by D.P. Forster and C.E.B. Frost taken from the *British Journal of General Practice*, **41** (1991), 67–71.

SUMMARY. There are proposals to set up prescribing budgets for family practitioner committees (now family health services authorities) and indicative prescribing amounts for practices. An intelligible model is therefore required for specifying budgetary allocations. Regression analyses were used to explain the variation in prescription rates and costs between the 98 family practitioner committees of England and Wales in 1987. Fifty-one per cent of the variation in prescription rates and 44% of the variation in prescription costs per patient could be explained by variations in the age–sex structure of family practitioner committees. The standardized mortality ratio for all causes and patients in 1987, and the number of general practice principals per 1000 population in 1987, but not the Jarman under-privileged area score were found to improve the predictive power of the regression models significantly ($p < 0.01$). The predictions of the model for the 10 family practitioner committees with the highest and lowest prescription rates or costs are reported and discussed. Potential improvements in models of prescribing behaviour may be thwarted by two problems. First, the paucity of readily available data on health care need at family practitioner committee and practice levels, and secondly, the increasing complexity in the statistical techniques required may render the procedure less intelligible, meaningful and negotiable in a contentious field.

## INTRODUCTION

Among the many proposals for the reform of the National Health Service are prescribing budgets for family practitioner committees (now family health services authorities) and indicative prescribing budgets for general practices.[1] There has been debate as to whether these proposed budgets imply cash limits, thus raising the issue of whether doctors will have to consider even more carefully the prescription of effective yet costly drugs for their patients. Moreover, there has been a concern that the proposals may provide a disincentive for doctors to screen patients for diseases such as hypertension, which, if detected, could lead to drug treatment.[2] The response of the Department of Health has been to soften the language used and the word 'budget' is employed only in the context of the indicative prescribing scheme for family practitioner committees and the word 'amount' is used instead for practices.[3]

Putting aside the medico-political debate concerning the advisability or otherwise of such budgets, a key question is whether data and methods exist to specify prescribing budgets for family practitioner committees and the equivalent for practices. Unless one accepts the view that each general practitioner throughout the country should be issuing the same number of prescriptions per patient or have identical prescribing costs per patient, a model is required linking the factors most likely to account for variations in prescribing behaviour with observed variations in such behaviour.

The purpose of this study was to explore one method of predicting or explaining observed variations in prescription rates and costs between family practitioner committees in terms of 'need for health care', including age–sex structure, and indicators of resource availability.

## METHOD

All data were collected for the 98 family practitioner committees of England and Wales. The total number of prescriptions in all therapeutic classes and the total costs of prescriptions for 1987 were derived from a representative one in 200 sample of prescriptions written by general practitioners and dispensed by retail pharmacies.[4] Similar prescription and cost data from dispensing general practitioners were collated and added to those dispensed by retail pharmacies.[4] The population data used were the Office of Population Censuses and Surveys mid-year estimates for each family practitioner committee in 1987.[5]

Three indicators of 'need for health care' were considered: the standardized mortality ratio for all causes and patients in 1987 (compares each family practitioner committee with a baseline of 100 for England and Wales as a whole); the Jarman under-privileged area score; and the age–sex structure of family practitioner committees in 1987. The Jarman score is a composite indicator based on eight social and demographic factors weighted by the importance placed on them for determining workload by a random sample of general practitioners.[6] The Jarman score is derived from data in the last decennial census in 1981. The indicator of resource availability considered was the supply rate of general practitioners in terms of the number of general practice principals per 1000 population in 1987.[7]

Ordinary least squares linear regression analysis was used to predict the number of prescriptions per patient for each of the 98 family practitioner committees, that is to say, regression analysis was used to establish the nature of the relationship between two or more variables.[8] In this study, there are two or more sets of values (for example, the prescription rate, the standardized mortality ratio and other demographic variables) and the value of the prescription rate that would correspond with given values of the standardized mortality ratio and other variables is predicted. The accuracy of the prediction will depend upon the strength of the relationship. This strength is usually measured in terms of the $\bar{R}^2$ statistic with a maximum value of 1.0 indicating 'a very strong relationship' and 0.0 indicating 'no discernable relationship'. In practice, less extreme values are frequently encountered. The same analytical method was used to estimate the value of total prescription costs per person (the predicted variable) given the values of the predictor variables (for example, the standardized mortality ratio and demographic variables).

## RESULTS

Figure 4.1(a) summarizes the results when simple linear regression was carried out using the standardized mortality ratio as the predictor variable and the prescription rate in 1987 as the variable to be predicted. In this analysis 46% of the variation in prescription rates between the family practitioner committees was explained by the variation in standardized mortality ratio ($\bar{R}^2 = 0.46$, $p < 0.001$). A similar procedure was used with the Jarman score as the predictor variable and in this instance 16% of the variation in the prescription rate was explained by the variation in the Jarman score ($\bar{R}^2 = 0.16$, $p < 0.001$) (Figure 4.1(b)).

These preliminary regressions suggest that it might be possible to model prescribing behaviour at family practitioner committee level, but they also indicate that there is considerable room for improvement. One plausible interpretation of the standardized mortality ratio regression is that there is some combination of age–sex effects in operation, and that such effects might be better investigated separately. Hence the regression results reported in the Appendix distinguish demographic factors from the remaining indicators of 'need for health care'. A selective summary of these results is reported in Table 4.1. From Table 4.1 it may be inferred that a superior model of prescribing behaviour will involve demographic factors in addition to the other two indicators of 'need for health care'. Overall, 51% of the variation in the prescription rate between the 98 family practitioner committees was explained by demographic factors alone. The addition of the standardized mortality ratio, the Jarman score and supply rate of general practitioners improved the predictive power of the model to 65%. No obvious advantage, however, was conferred upon the model by the inclusion of the Jarman score in the set of predictor variables (Appendix). In a second set of regression models, the same approach was used to explain the variation in prescribing costs per patient. In this case 44% of the variation in prescribing costs between the 98 family practitioner committees was ex-

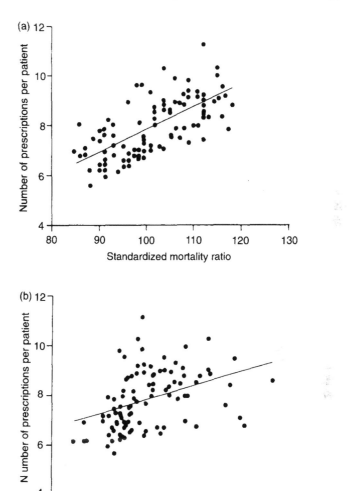

**Figure 4.1** Prescription rate for each family practitioner committee in 1987 against (a) standardized mortality ratio and (b) Jarman under-privileged area score

plained by demographic variables alone. The inclusion of the standardized mortality ratio, Jarman score and supply rate of general practitioners offered a considerable improvement in that 60% of the overall variation was explicable. Once again the

**Table 4.1** Prediction of prescription rates and costs for all therapeutic classes

| | $\bar{R}^{2}*$ for: | |
| | No. of prescriptions per patient | Cost of prescriptions per patient |
| Predictor variables | | |
|---|---|---|
| *Simple regressions* | | |
| Standardized mortality ratio | 0.46 | N/A |
| Jarman score | 0.16 | N/A |
| *Phase I regressions* | | |
| Age–sex structure | 0.51 | 0.44 |
| *Phase II regressions* | | |
| Phase I plus SMR, Jarman score, ro. of GPs per 1000 population | 0.65 | 0.60 |

* $\bar{R}^{2}$ Represents the proportion of the variation in prescription rates between the 98 family practitioner committees which is accounted for by the predictor variable or variables. N/A = not applicable. SMR = standardized mortality ratio.

inclusion of the Jarman score in the group of predictor variables conferred no clear advantage (Appendix).

Table 4.2 sets out the 10 family practitioner committees with the highest and lowest actual prescribing rates in 1987. These ranged from 11.1 prescriptions per person in Mid Glamorgan to 5.7 in Oxfordshire. These observed prescribing rates are contrasted with the rates predicted by the regression model described in the Appendix (phase II model). Table 4.2 also sets out the extremes in prescribing costs per patient in 1987. These varied from £54 per person in Mid Glamorgan to £35 in Oxfordshire, although the relative positions of other family practitioner committees for actual prescription rates and costs were not identical. The overall figures for England and Wales are also given in Table 4.2. It is important to note that for the family practitioner committees with the highest actual prescribing rates and costs, the predicted value tends to underestimate the actual value's departure from the implied regression line, and vice versa for the family practitioner committees with the lowest rates and costs. The spread of the differences between the actual and predicted prescription rates was more evenly distributed for the family practitioner committees with less extreme rates and costs.

**Table 4.2** Actual and predicted prescribing rates and costs in all therapeutic classes for the ten family practitioner committees with the highest and lowest actual values in 1987

| Family practitioner committee | No. of prescriptions per patient | | Family practitioner committee | Cost of prescriptions per patient (£) | |
|---|---|---|---|---|---|
| | Actual | Predicted* | | Actual | Predicted* |
| Mid Glamorgan | 11.1 | 9.2 | Mid Glamorgan | 54.1 | 48.0 |
| West Glamorgan | 10.4 | 8.9 | North Tyneside | 52.9 | 50.1 |
| Liverpool | 10.4 | 9.8 | Salford | 52.7 | 49.5 |
| Salford | 10.1 | 9.7 | Gwynedd | 51.9 | 47.8 |
| Gwent | 9.9 | 8.5 | West Glamorgan | 51.7 | 47.1 |
| Barnsley | 9.8 | 8.1 | Gwent | 49.8 | 45.2 |
| Trafford | 9.6 | 7.7 | Liverpool | 49.6 | 49.7 |
| Gwynedd | 9.6 | 9.0 | Barnsley | 49.4 | 43.9 |
| Manchester | 9.5 | 9.2 | Clwyd | 49.2 | 44.2 |
| Sandwell | 9.3 | 9.5 | Trafford | 48.9 | 42.4 |
| Barnet | 6.5 | 6.7 | Hertfordshire | 37.0 | 39.5 |
| Gloucestershire | 6.5 | 7.3 | Wiltshire | 36.7 | 38.7 |
| Croydon | 6.4 | 6.7 | Enfield and Haringey | 36.5- | 39.8 |
| Wiltshire | 6.4 | 6.9 | Berkshire | 36.4 | 36.6 |
| Hertfordshire | 6.3 | 7.1 | Gloucestershire | 36.2 | 41.7 |

## DISCUSSION

The approach to explaining the variation in prescription rates used in this study has been to accept that general practitioners' responses to morbidity in their practices will, in part, be met by prescriptions. Thus, explanations of prescribing behaviour are only understandable and likely to be successful if the level of morbidity in a practice or family practitioner committee area can be measured. An indicator of 'need for health care', or more accurately 'need for a prescription' is elusive. There is a paucity of useful sources of meaningful morbidity data at general practice or family practitioner committee level. For example, continuous morbidity recording in general practice is carried out by relatively few general practitioners.[9]

For this reason, three proxy indicators of the need for health care were considered: the standardized mortality ratio, the Jarman score and demographic factors, that is, age–sex structure. The all-causes standardized mortality ratio, used first, may appear inappropriate since much general practice work does not involve life threatening conditions.[10] It does, however, have the advantage of being readily available each year at family practitioner committee level and could be obtained at individual practice level on a regular basis, probably by aggregating data for several years and using, say, a five year rolling average. It is noteworthy that, even after allowing for the effect of the demographic factors, the standardized mortality ratio was found to exert a positive influence on the prescription rate and costs.

The second indicator of health care need used was the Jarman under-privileged area score derived from 1981 census data. This indicator is intended to convey the pressure of work on general practitioners. Given that the eight elements making up the Jarman score are weighted by the subjective views of general practitioners with respect to the pressure of work generated, it was anticipated that the Jarman score might give a good explanation of the variation in prescription rates. Regression equations using the Jarman score as a predictor variable, however, gave a relatively poor explanation of the variation in prescription rates or costs. Furthermore, the inclusion of the Jarman score along with other indicators of 'need for health care' did not improve the explanatory power of the models. It seems likely that the Jarman score, based essentially on 1981 data, may be inapplicable for use in years distant from the date of the census. Moreover, the Jarman score would be impossible to calculate accurately from census data for individual practices.[11,12]

In the Appendix, variation in resource availability in the form of the number of general practitioners per 1000 population was included in the regression models for the following reasons. In most studies of the use of health services, it has been found that increased availability of supply increases health service utilization.[13,14] Elsewhere, it has been argued that the availability of good medical care varies inversely with the need for health care.[15] Apparently, the supply rate of general practitioners does offer a worthwhile improvement when employed as a predictor variable. In addition, the supply rate appears to be positively related to the prescription rate and costs. This finding appears to corroborate the previously noted utilization effects and is not inconsistent with the 'inverse care' law.[15]

For all practical purposes a regression model explaining about 65% of the observed variation in prescription rates and 60% in the case of costs is available if the standardized mortality ratio and supply rate of general practitioners are combined with demographic measures. Ideally, it would be preferable to specify a regression model explaining a much higher percentage of the observed variation in prescription rates and costs. There are a number of ways in which such an improvement might be attained. First, additional predictor variables could be introduced into the linear regression equations. Secondly, it may be possible to improve the explanation of prescribing behaviour by using different regression analysis techniques. Briefly, these might include weighted least squares regression or non-linear least squares regression methods.

The government intends that family practitioner committees with above average actual prescribing costs should bring their expenditure into line with lower spending family practitioner committees. It is accepted, however, that family practitioner committees may have high prescribing costs because of special known factors. In our analysis, we have interpreted these special factors principally as 'need for health care' including demographic factors. We have introduced a reasonably satisfactory model of prescribing behaviour, which is note-worthy for its reliance on demographic factors, the standardized mortality ratio and the supply rate of general practitioners. One practical implication of this model is that family prac-titioner committees with high prescribing costs per patient, such as Mid Glamorgan or North Tyneside, would come under less financial pressure than if they were compared with some nationally derived average. In practice, since family practitioner committee budgets will probably be decided at regional level, each family practitioner committee's actual prescription rate and costs will be compared with the appropriate regional average. Whether there would be concomitant generosity in the allocation of prescribing budgets to family practitioner com-mittees in the south, for example, Buckinghamshire and Oxfordshire, which are underspending relative to the model's predictions, is a moot point.

In family practitioner committees in which prescription rates or prescribing costs are in excess of those predicted by the model, two explanations are possible. It may be that prescribing is unjustifiably high, or that the high prescription rate or costs reflect factors not contained in the model. We favour the latter explanation and have noted that routinely available data do not exist to improve the predictive power of the model. We have indicated that there are a number of technical and statistical improvements that might be made to the model presented here, but at some risk of increasing its overall complexity. In which case the model of prescribing behaviour becomes less understandable, meaningful and negotiable. This bodes ill in a field in which the ideology of budgets has not generally been well received.

While we agree with the government's concentration on age and sex structure in setting family practitioner committee budgets and indicative prescribing amounts in practices, our view is that the scheme is premature, given that some 49% of prescribing variation is unexplained.

## APPENDIX

The regressions reported here were undertaken in two phases. From studies in selected practices, it is known that morbidity rates and prescription rates vary in different age and sex groups, and are particularly high in the young and elderly age groups.[9] Using available demographic data about the proportions of a family practitioner committee's population in the young or elderly age bands and the reproductive age group for women, maximum prediction of the observed variation in prescription rates or costs was sought from the phase I regressions. More precisely, in statistical terms, the square of the product-moment correlation between the dependent variate and the set of independent variates, known as the coefficient of determination, adjusted for degrees of freedom, that is $\bar{R}^2$, was maximized.[16]

At the conclusion of phase I, the overall variation in the prescription rate or costs was divided into two parts: the part predicted or explained by the age and sex distributions in the family practitioner committees and the age and sex distributions in the family practitioner committees and the part remaining to be explained (the residual variation). In phase II, therefore, other potentially relevant indicators, for example the Jarman score and standardized mortality ratio, were introduced into the regression analysis to see if their effects on prescription rates or costs were statistically significant. In addition, the resource availability indicator, the supply rate of general practitioners, was introduced at this stage.

Table 4.3 summarizes the phase I results for prescription rates for all therapeutic classes, and for prescribing costs. In the usual case, the regression coefficient of a predictor variable measures the effect on the predicted variable of a unit increase in the predictor variable, holding any other predictor variables constant. When regressions involve proportions, the interpretation given to the regression coefficients must be amended because the proportion of children under one year of age cannot be increased, while at the same time holding the remaining age–sex proportions constant. In these regressions, the coefficients of the regression may be related to the prescription rate for a particular age–sex group.[16] For example, for the prescription rate for children aged under one year:

Regression coefficient of constant + regression coefficient

of children aged under one year = 6.5 + 383.9 = 309.4

This interpretation immediately raised a question as to why certain coefficients were statistically significant and yet had

**Table 4.3** Phase I regressions: explanation of the variation in prescription rates and prescribing costs for all therapeutic classes (dependent variables) by the age–sex structure of the 98 family practitioner committees in 1987

| Label | Partial regression coefficient | t | Significance level |
|---|---|---|---|
| *No. of prescriptions per patient** | | | |
| Constant | 6.45 | 1.45 | NS |
| Proportion of population: | | | |
| Aged under 1 year | 383.85 | 2.37 | $p < 0.05$ |
| Aged 1–4 years | −67.91 | −1.32 | NS |
| Males aged 75 years and over | −291.97 | −7.15 | $p < 0.001$ |
| Females aged 65–74 years | 207.92 | 6.03 | $p < 0.001$ |
| Females aged 15–44 years | −18.18 | −1.30 | NS |
| *Prescribing costs per patient*** | | | |
| Constant | 20.31 | 6.80 | $p < 0.001$ |
| Proportion of population: | | | |
| Males aged 75 years and over | −1047.49 | −6.54 | $p < 0.001$ |
| Females aged 65–74 years | 923.42 | 8.79 | $p < 0.001$ |

In each of the regressions reported above, $\bar{R}^2$ was maximized using various elements of family practitioner committee age–sex structure. NS = not significant. *92 degrees of freedom. $\bar{R}^2 = 0.51$. **95 degrees of freedom. $\bar{R}^2 = 0.44$.

unrealistically large or even negative values. There probably exists a strong linear relation connecting many, if not all, the predictor variables. The predictor variables are then said to be collinear and the coefficients of the regression become indeterminate with occasionally, but not inevitably, large standard errors.[16]

Age–sex standardized prescription rates and costs were required for phase II regressions involving the standardized mortality ratio and possibly the Jarman score. The phase I regressions provided a means of obtaining such data indirectly, and thus the collinearity problem did not matter given the strong relationship between the predicted variable and the set of phase I predictors.

The effects of the remaining 'need for care' and resource availability indicators on prescription rates and the attendant

**Table 4.4** Phase II regressions: explanation of the variation in prescription rates and prescribing costs for all therapeutic classes (dependent variables) by the indicators of 'need for health care' of family practitioner committees in 1987

| Label | Partial regression coefficient | t | Significance level |
|---|---|---|---|
| *No. of prescriptions per patient** | | | |
| Constant | 2.63 | 0.52 | NS |
| Proportion of population: | | | |
| Aged under 1 year | 67.52 | 0.43 | NS |
| Aged 1–4 years | −9.07 | −0.20 | NS |
| Males aged 75 years and over | −158.98 | −2.95 | $p < 0.01$ |
| Females aged 65–74 years | 115.94 | 3.37 | $p < 0.01$ |
| Females aged 15–44 years | −26.74 | −2.05 | $p < 0.05$ |
| No. of GPs per 1000 population | 6.28 | 2.95 | $p < 0.01$ |
| Jarman score | 0.01 | 1.46 | NS |
| SMR (all causes and patients) | 0.05 | 3.17 | $p < 0.01$ |
| *Prescribing costs per patient*** | | | |
| Constant | −21.06 | −2.46 | $p < 0.05$ |
| Proportion of population: | | | |
| Males aged 75 years and over | −432.84 | −1.90 | NS |
| Females aged 65–74 years | 586.64 | 4.51 | $p < 0.001$ |
| No. of GPs per 1000 population | 33.85 | 3.86 | $p < 0.001$ |
| Jarman score | −0.003 | −0.11 | NS |
| SMR (all causes and patients) | 0.26 | 4.08 | $p < 0.001$ |

In each of the regressions reported above, $\bar{R}^2$ was initially maximized using family practitioner committee age–sex structure (see Table 4.3). The need for health care indicators including the supply rate of general practitioners were subsequently added. NS = not significant. SMR = standardized mortality ratio. * 89 degrees of freedom. $\bar{R}^2 = 0.65$. ** 92 degrees of freedom. $\bar{R}^2 = 0.65$.

prescribing costs were explored by a number of regressions. In each case the conclusion was virtually the same, the additional explanatory power offered by the inclusion of these variables, when added separately or together, was substantial except in the case of the Jarman score. A representative set of these phase II results is given in Table 4.4. Although the standardized mortality ratio and the Jarman score were apparently collinear, their inclusion in the same regression equation did not materially affect the phase II results reported here.

REFERENCES

1. Secretaries of State for Health, Wales, Northern Ireland and Scotland. *Working for patients: indicative prescribing budgets for general medical practitioners.* NHS review working paper 4. London: HMSO, 1989.
2. Spencer, J.A. and van Zwanenberg, T.D. Prescribing research: PACT to the future. *J R Coll Gen Pract* 1989, **39**: 270–272.
3. Department of Health. *Improving prescribing. The implementation of the GP indicative prescribing scheme.* London: Department of Health, 1990.
4. Prescription Pricing Authority. *Statistical data relating to prescriptions dispensed by chemists, contractors and dispensing doctors for 1987. Forms PD1 and FP51.* London: Department of Health, 1987.
5. Office of Population Censuses and Surveys. *Key population and vital statistics. England and Wales for 1987. Series VS No. 14/PPI No. 10.* London: HMSO, 1989.
6. Jarman, B. Identification of underprivileged areas. *Br Med J* 1983, **286**: 1705–1709.
7. General Medical Services. *Basic statistics for 1987.* London: Department of Health, 1987.
8. Godfrey, K. Statistics in practice: simple linear regression in medical research. *N Engl J Med* 1985, **313**: 1629–1636.
9. Royal College of General Practitioners, Office of Population Censuses and Surveys and Department of Health and Social Security. *Morbidity statistics from general practice. Third national study, 1981–82.* London: HMSO, 1986.
10. Fry, J. *Common diseases.* Lancaster: MTP Press, 1985.
11. Hutchinson, A., Foy, C. and Smyth, J. Providing census data for general practice. 1. Feasibility. *J R Coll Gen Pract* 1987, **37**: 448–450.
12. Foy, C., Hutchinson, A. and Smyth, J. Providing census data for general practice. 2. Usefulness. *J R Coll Gen Pract* 1987, **37**: 451–454.
13. Feldstein, M.S. Effects of differences in hospital bed scarcity on type of use. *Br Med J* 1964, **2**: 562–565.
14. Cullis, J.G., Forster, D.P. and Frost, C.E.B. The demand for inpatient treatment: some recent evidence. *Applied Economics* 1980, **12**: 43–60.
15. Hart, J.T. The inverse care law. *Lancet* 1971, **1**: 405–412.
16. Maddala, G.S. *Econometrics.* New York: McGraw-Hill, 1977.

## 4.1 DISCUSSION

### 4.1.1 Objective

The purpose of the study was to explore one method of predicting or 'explaining' observed variations in prescription rates and costs between family practitioner committees in terms of 'need for health care', including age–sex structure and indicators of resource availability.

### 4.1.2 Design

A historical survey was made of the prescription rates per patient, and of the prescription costs per patient, in both cases for the 98 family practitioner committees of England and Wales in 1987. The variation in these two 'dependent' variables was thought to be 'explainable' by a number of 'independent' or 'explanatory' variables which were also collected for each of the 98 committees in 1987. These independent variables consisted of several describing the age–sex structure of the patients, the standardized mortality ratio (SMR) for all causes and patients, the Jarman under-privileged area score and, finally, the number of general practitioner principals per 1000 population.

The aim of the statistical analysis of the data was to relate each dependent variable to one or more of the independent (explanatory) variables. The statistical technique used is called linear regression analysis, which is referred to as **simple** linear regression analysis when there is only one independent variable, and **multiple** linear regression analysis when there are two or more independent variables. The outcome of the analysis is an equation or 'model' containing only those independent variables which explain a significant amount of the variation in the dependent variable.

### 4.1.3 Subjects

The 'subjects' were the 98 groups of patients corresponding to 98 family practitioner committees of England and Wales in 1987.

### 4.1.4. Outcome measures

There are two dependent variables: prescription rates per patient; and prescription costs per patient.

Several of the independent (explanatory) variables are concerned with describing the age–sex structure of the patients in terms of proportions of population: those aged under 1 year; those aged 1–4 years; males ages 75 and over; females aged

*Prescribing rates and costs*

65–74; and females aged 15–44. Other independent variables are the standardized mortality ratio (SMR); Jarman score; and the number of GPs per 1000 population.

### 4.1.5 Data and statistical analysis

Data were collected for all the variables listed in section 4.1.4 for each of the 98 family practitioner committees of England and Wales in 1987. Each dependent variable was related to the independent variables using multiple linear regression analysis. However the Results section of the case study begins with a simple linear regression analysis, namely one relating the dependent variable 'prescription rate per patient' to the independent variable 'SMR'. Section 12.1 discusses the concepts and methods of simple linear regression analysis, and uses the application in this case study to illustrate the methods. To keep computations simple, only a random sample of ten of the 98 family practitioner committees data has been included.

Having carried out some preliminary regression analysis using first SMR, and then Jarman score to 'explain' the variation in prescription rate per patient, the Results section goes on to discuss multiple linear regression, that is, how other independent variables are introduced into the regression equation to try to build a 'better' model, i.e. one which explains more of the variation in the dependent variable as measured by $\bar{R}^2$. Also discussed are regressions in which the dependent variable is 'prescription costs per patient'.

In section 12.6 there is a brief discussion of the concepts and methods of multiple linear regression using the method of ANOVA first described in Chapter 9. It is not possible to give any examples of multiple regression in Chapter 12 using the actual data of case study 4 since these data are not available in sufficient detail. However a certain amount of fictitious data has been used in Chapter 12 to illustrate one of the difficulties of performing and interpreting the results of multiple linear analysis. Readers should also study the Appendix to case study 4 for further discussion of this type of analysis.

The conclusion reached in case study 4 is that models for prescription rates per patient should include a number of explanatory variables to boost the value of $\bar{R}^2$ to 0.65. Similar conclusions are drawn for models for prescription costs per patient for which a value of 0.60 for $\bar{R}^2$ is achievable. The Discussion section of case study 4 focuses on the practical benefits to GPs of these statistical models.

### 4.1.6 Further points

The question of 'study size' is relevant here, as are the concepts of 'population' and 'sample'. The justification for choosing the 98 family practitioner committees in England and Wales in 1987 was presumably that they make up all such committees and there were not too many of them to warrant some sampling procedure. This is fine, up to a point. But it does mean that inferences based on statistics for the 98 committees in 1987 intended presumably to apply to some larger 'population' have no meaning since data from the whole 'population of committees' have been analysed. It may be that the author of the case study assumed that the models derived from 1987 data apply to 1988 and subsequent data. This seems to be an assumption of doubtful validity.

# 5

# Hidden psychiatric illness: use of the general health questionnaire in general practice

*He [Lord Ellenborough] could not for the life of him see what
was the use of asking people so many questions. Here, then,
Miss Nightingale was in advance of her time; in one case, by a
generation, in the other, by two generations.*

*From a House of Lords debate on the Census of 1860.*

Case study 5 is an article by Alastair F. Wright and
Anthony F. Perini taken from the *Journal of the Royal
College of General Practitioners*, **37** (1987), 164–7.

SUMMARY. A 10% random sample comprising 234 adults attending a general practitioner was studied to obtain an estimate
of conspicuous and hidden psychiatric morbidity and to determine the value of the general health questionnaire in improving
case recognition in general practice. Patients completed the 28-
item general health questionnaire before seeing the general
practitioner, who completed a rating sheet without seeing
the general health questionnaire score. The doctor identified a
psychiatric component in 38% of men and 53% of women and
diagnosed psychiatric disorder in 22% of men and 31% of
women. Using a cut-off point of nine or above, high general

health questionnaire scores were found in 25% of men and 29% of women. Agreement between the general health questionnaire and the doctor's assessment was better for males (misclassification rate 16%) than for females (20%). A subsample of patients scoring over the recommended threshold (five or above) on the general health questionnaire were interviewed by the psychiatrist to compare the case detection of the general practitioner, an independent psychiatric assessment and the 28-item general health questionnaire at two different cut-off scores. The general health questionnaire may be a useful tool for improving recognition of psychiatric morbidity in general practice if the cut-off point is raised above that recommended for epidemiological research.

## INTRODUCTION

It is estimated that between 10% and 20% of the general practice population are mentally or emotionally disturbed[1] and the satisfactory management of patients with psychological symptoms is particularly difficult in general practice, given the constraints of limited consulting time. Continuing care of some such patients may be stressful and unrewarding for both doctor and patient in terms of the inability of the doctor to achieve a 'cure' or to intervene to the same extent as with major physical illness.

There is considerable evidence[2-5] to suggest that many patients with significant psychiatric illness attending their general practitioner are unrecognized as such. In 1970 Goldberg and Blackwell[2] coined the term 'hidden psychiatric morbidity' and found that these patients were distinguished by their attitude to their illness and by usually presenting a physical symptom to the general practitioner.

In 1984 Skuse and Williams[6] investigated psychiatric morbidity in a consecutive series of 303 patients attending an experienced south London general practitioner. The doctor classified 24% of these patients as psychiatric cases while the estimated true prevalence of psychiatric cases in the sample was 34%. The discrepancy was largely owing to the general practitioner regarding expected depressives as normal rather than as cases and giving them a different diagnosis.

The development of the general health questionnaire by

Goldberg[7] has been a significant advance in psychiatric epidemiology. The questionnaire may also have a value in assisting a doctor to identify psychiatric illness in his patients so that appropriate treatment may be begun. The general health questionnaire[8–12] is a self-reporting screening questionnaire which identifies individuals who have a high probability of suffering from psychological illness. It has been well validated and correlates well with the assessments of consultant psychiatrists. The scoring distinguishes between chronic stable complaints and recent exacerbations, an item being counted if the patient thinks there has been a change from his/her 'usual self' over the last week. The questionnaire is specifically designed for use in community settings.

The aims of this study were to obtain an estimate of conspicuous and hidden psychiatric morbidity in a general practice and to estimate the value of the general health questionnaire in improving case recognition in a clinical setting.

## METHOD

The study population was drawn from patients attending one general practitioner in a health centre group practice over a period of five months in 1985. A sample was selected so that all the consulting sessions in the week were represented by a 10% random sample of patients attending that session and no patient was included more than one. The sample comprised 234 patients who were over 17 years of age but under 65 years. House calls, which account for approximately 8% of the workload, were excluded.

All 234 adult patients were asked to complete the 28-item version of the general health questionnaire and to answer supplementary questions on social and employment status before seeing the doctor. Without seeing the general health question naire score the general practitioner completed a rating sheet after seeing each patient. The rating sheet consisted of three assessments: (1) the reason for consultation using the categories of Goldberg and Blackwell; (2) the degree of psychiatric disorder, on a five-point scale; (3) the diagnosis (up to two items) and a list of known chronic conditions (up to four).

A 20% random subsample of respondents scoring five or over on the 28-item general health questionnaire were seen by the psychiatrist who completed the general practice research unit's clinical interview schedule.[7,13] The clinical interview schedule is a semi-structured interview intended for use in the community by trained psychiatrists. Section 1 includes 10 symptoms, each rated 0–4. A morbid score (2, 3 or 4) in this section defines a 'case'. The clinical interview schedule concentrates mainly on neurotic symptoms but the psychiatrist is free to explore the possibility of psychosis or organic impairment as in a normal clinical interview. The sum of section 1 and section 2 scores gives an overall severity score. In addition, where appropriate, the psychiatrist records a diagnosis using the ninth revision of the *International classification of diseases*.[14]

Because of limited resources and the time required for psychiatric interviews, no attempt was made to validate the general health questionnaire in this population against the psychiatrist's diagnosis. Instead, the psychiatric interviews were used to provide a limited comparison between the case detection of the general practitioner, an independent psychiatric assessment and the general health questionnaire at two different cut-off points.

RESULTS

All 234 general health questionnaires were returned but seven were rejected for analysis. One had not been filled in as the patient spoke only Mandarin Chinese, and six had many missing answers. Twenty-four questionnaires had one or two answers missing but these appeared to have been overlooked and not avoided intentionally. Therefore 227 questionnaires (97%) were analysed – 115 from men and 112 from women.

Sixty-nine per cent of the men and 71% of women classified themselves as married and 13% and 14%, respectively, as divorced, separated or widowed. Five per cent of men and 4% of women classified themselves as unemployed, while 5% of husbands or male partners were unemployed. Past history of psychiatric illness was admitted by 54% of women and 31% of men; 25% of women and 8% of men had been treated for this in the previous year.

Diagnoses made by the general practitioner were classified according to the classification of the Royal College of General Practitioners.[15] The most common primary reason for consulting in female patients was psychiatric illness (19%), followed by maternity services (16%). Women who made an appointment for a child in order to consult indirectly about their own emotional problems were not counted in the total for psychiatric illness. For men, respiratory illness was the most common reason for consulting (15%), while psychiatric illness was in fourth position.

The general practitioner's assessment of the type of disturbance is given in Table 5.1. Taking categories 2 to 6 inclusive as indicating a psychiatric component to the consultation, the doctor identified a psychiatric component in 38% of men and 53% of women.

The doctor also assessed the degree of disturbance, identifying as cases those patients who had distressing psychological symptoms and also some disturbance of normal social functioning. Clinical psychiatric illness was diagnosed by the doctor in 22% of men and 31% of women. While mild psychiatric disturbance was more common in women, the percentage of moderate and severe psychiatric illness was the same (12%) in both sexes.

**Table 5.1** The general practitioner's assessment of the type of disturbance suffered by respondents

| Category | Number (%) of respondents | | | |
|---|---|---|---|---|
| | Men | | Women | |
| 1. Entirely physical complaint | 68 | (59) | 28 | (25) |
| 2. Physical complaint in neurotic person | 8 | (7) | 19 | (17) |
| 3. Physical complaint plus associated psychiatric illness | 17 | (15) | 17 | (15) |
| 4. Psychiatric illness plus somatic symptoms | 0 | (0) | 4 | (4) |
| 5. Unrelated physical and psychiatric complaints | 7 | (6) | 5 | (4) |
| 6. Entirely psychiatric illness | 11 | (10) | 15 | (13) |
| 7. Not ill, unclassifiable | 4 | (3) | 24* | (21) |
| Total | 115 | (100) | 112 | (100) |

* Includes 12 antenatal patients.

## Misclassified patients using a threshold score of nine

The manual for the 28-item general health questionnaire[16] recommends a threshold score of five, threshold being defined as 'just significant clinical disturbance' or that point where the probability of being a 'case' is 50%. In general practice it would be more clinically useful for the test to have high specificity (that is, relatively few false positives), thus excluding patients whose symptoms are so mild that no therapeutic action is called for.

The general health questionnaire results obtained here have been used as the standard against which the doctor's assessment is compared and as the basis of calculations of sensitivity and specificity for different threshold scores. Based on these calculations (Figure 5.1) it is suggested that a cut-off point of nine is best suited for clinical use in general practice.

Using this cut-off point high scores on the general health questionnaire were found for 25% of men and 29% of women. Misclassification of patients is given in Table 5.2 and a definition of the statistics used for the analysis is also given. Correlation between the general health questionnaire and the doctor's assessment was better for men than for women. The overall

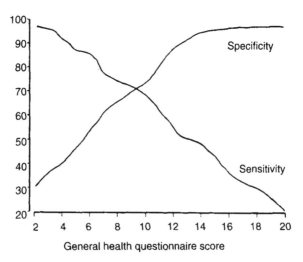

**Figure 5.1** Sensitivity and specificity of the general health questionnaire versus the general practitioner's assessment

**Table 5.2** Comparison of the doctor's assessment and the general health questionnaire (GHQ) score for male and female respondents using a threshold score of nine

| | Doctor's assessment | | | | | |
| | Number of men | | | Number of women | | |
| GHQ score | Not a psychiatric case | Psychiatric case | Total | Not a psychiatric case | Psychiatric case | Total |
|---|---|---|---|---|---|---|
| <9 | 79 | 7 | 86 | 67 | 12 | 79 |
| ⩾9 | 11 | 18 | 29 | 10 | 23 | 33 |
| Total | 90 | 25 | 115 | 77 | 35 | 112 |

$$\text{Misclassification rate} = \frac{\text{false negatives} + \text{false positives}}{\text{grand total}} \times 100$$

$$\text{Specificity} = \frac{\text{true negatives}}{\text{true negatives} + \text{false positives}} \times 100$$

$$\text{Sensitivity} = \frac{\text{true positives}}{\text{true positives} + \text{false negatives}} \times 100$$

$$\text{Positive predictive value} = \frac{\text{true positives}}{\text{false positives} + \text{true positives}} \times 100$$

misclassification rate was 16% for men and 20% for women. Similarly specificity was high at 88% and 87%, respectively, and sensitivity was acceptable at 72% and 66%, respectively. The positive predictive value, that is the probability of a high score being a case, was 62% and 70%, respectively.

## Results of psychiatric interviews

Nineteen patients were referred for standardized psychiatric interview.[13] Two male patients failed to attend – 10 women and seven men were interviewed. Some patients were seen the same day and most within a week of seeing the general practitioner.

Three female patients regarded as cases by the doctor were not judged as cases by the clinical interview schedule. In two of these patients, the doctor's background knowledge may have biased judgement in favour of a psychiatric condition. One male patient, not regarded by the doctor as a case, was regarded as a case by the clinical interview schedule. This man's long history of physical illness and tendency to complain may also have influenced the doctor's judgement. The general health questionnaire scores for these patients were in the range 10–13.

Using a cut-off score of five on the general health questionnaire (17 patients, a 20% subsample) the clinical interview schedule detects 10 cases (59%) and the doctor 12 (71%). However using a cut-off score of nine (14 patients, no longer a random sample) the clinical interview schedule detects 10 cases (71%) and the doctor nine (64%).

## Clinical implications

The doctor took the opportunity to review his diagnosis retrospectively when his assessment did not agree with the general health questionnaire score.

Seven men with low scores had been classified as cases by the doctor. Four had a diagnosis of depression (taking high dose antidepressant drugs) or anxiety state, while one other had a past history of recurring psychosis. One of the remaining two patients had Paget's disease of bone and Parkinson's disease, and the other had dyspepsia and long-standing marital problems. While the doctor may have been unduly influenced by his personal knowledge of the patients and their psychiatric histories, it seems more likely that the discrepancy was due to the general health questionnaire failing to identify chronic stable complaints, for which it was not designed. Similarly, 12 women with low scores had been classified as cases by the doctor. One patient was a delusional schizophrenic, four had anxiety states and four others had 'panic attacks', 'tension headaches', 'personality disorder' and 'unexplained abdominal pains'. The remaining three had come respectively because of coryza, dyspepsia and a recent sterilization operation.

More important clinically are the 11 men and 10 women identified as cases by the general health questionnaire and missed by the doctor. Five of the men had serious chronic physical illness, one had very bad psoriasis and one (with a score of 19) had a long history of unexplained abdominal pain. Of the four men without chronic conditions one had a recent myocardial infarct, one a pilonidal sinus and the other two apparently minor infections. The women had lower scores than the men and six had chronic physical illnesses. Of the other four, two had migraine and the others were seen for infertility and a doubtful cervical smear. None of these patients, male or female, was known to suffer from chronic or recurrent psychiatric illness.

## DISCUSSION

In the UK, the great majority of patients with psychological symptoms are treated solely by their general practitioner. It is therefore pertinent to ask how common is psychiatric illness in patients attending their general practitioner and what can be done to improve recognition by the general practitioner.

The first aim of this study was to obtain an estimate of conspicuous and hidden psychiatric morbidity. There are major difficulties in defining what doctors mean by 'psychiatric disorder', in particular in limiting the term so that it does not become so broad as to be meaningless. The identification of a case by the general practitioner depended on the presence of distressing psychological symptoms and some disturbance of normal social functioning.

The second aim was to estimate the value of the general health questionnaire to the general practitioner in clinical practice rather than to validate the instrument in this population against psychiatric interview. It proved impracticable to arrange psychiatric interview of a large enough sample of all attending patients so our limited resources were directed to a comparison of the case-detecting behaviour of the doctor (who was actively looking for psychiatric disturbance) and independent psychiatric assessment.

Using the psychiatric interview as the ultimate criterion in diagnosing psychiatric illness, Skuse and Williams[6] pointed out that the general health questionnaire tended to over-identify and the general practitioner under-identify, while Goldberg and Bridges[17] showed that there is not complete agreement between psychiatric research interviews about what is thought to constitute a psychiatric case. General practitioners will have their own views and it may be that the doctor who is seeing eight patients in an hour finds it more profitable to make a prognostic assessment based on the patient's personality, social functioning and previous illness behaviour, whereas the psychiatric interview places more emphasis on diagnosis and psychiatric symptomatology.

The results found in this study are not dissimilar to those found in other published work.[2,18] Hoeper and colleagues[19] found that 27% of patients attending their general practitioner had psychiatric illness and 30% had high general health questionnaire scores. Goldberg and Blackwell[2] estimated that hidden psychiatric morbidity accounted for one third of all disturbed patients.

Goldberg[20] suggests that 'When a patient is found to have a high score the most natural response by the clinician is to look at the questionnaire with the patient and ask additional probe questions suggested by particular symptoms'. This course was followed by the general practitioner, but it was obvious early in the study that using the recommended threshold score of five seemed to produce more false positives than true positives. Raising the cut-off point to nine seemed to make the test more useful clinically. A similar conclusion is reported by Nott and Cutts[21] who studied 200 post-partum women from five Southampton general practices. They conclude that 'Slight modification of the content and a raised cut-off point of the general health questionnaire-30 make it a useful screening instrument for post-partum psychiatric disorder.'

The threshold score of nine or above seemed to give the best trade-off between sensitivity and specificity: the inevitable loss of sensitivity may not be important clinically as many low scorers may have only mild or transient disturbances.[17] Ultimately, the threshold score used is a matter of judgement, a compromise between cost and benefits and this compromise may vary from practice to practice. We suggest that a cut-off score of nine is a useful starting threshold which can be 'fine-tuned' on implementation in a particular practice.

Goldberg and Bridges[17] have shown that the use of the general health questionnaire by general practitioners could improve their ability to recognize hidden psychiatric morbidity in new episodes of illness. We have highlighted a clinically important group with chronic illness, whose physical disease was well known to the doctor, but whose psychiatric disturbance went unrecognized. It seems that the doctor's personal knowledge of his patient, normally so valuable in managing illness, may be a two-edged sword making the recognition of concurrent emotional illness less likely. In these circumstances the general health questionnaire would seem to have an important role in alerting to doctor.

The diagnosis and management of psychiatric illness in general practice presents formidable practical problems to the practitioner, particularly when seeing eight patients per hour. Assessment by standardized clinical interview schedule is very demanding of the psychiatrist's time and referral to a psychiatrist, for validation of the general practitioner's opinion is not always clinically desirable. A simple test, a kind of 'psychiatric ESR', would be invaluable. When the test result differed from clinical opinion then it would be a warning to see the patient again. Clinicians are accustomed to using biological tests as diagnostic aids and, while these estimations are subject to various errors and require clinical interpretation, they are more readily accepted as useful than are the screening tests familiar to consultant psychiatrists.

Experience from this study suggests that the general health questionnaire is simple to use in general practice and may prove useful in assessing patients with physical symptoms not conforming to any recognizable clinical pattern, and also frequent attenders. It may uncover unsuspected psychiatric illness, particularly in patients with chronic physical disease. It may even have a use in following progress in chronic psychiatric illness using a different scoring method as suggested by Goodchild and Duncan-Jones[22] though this has not been tested in the present work.

It is as yet uncertain whether improved recognition of psychiatric illness would necessarily improve treatment. Indeed, Freeling and colleagues[23] studied unrecognized depressives and concluded that they seemed to suffer rather than benefit from continuing care. What is clear is the need for further systematic study of these disorders in general practice. The patients in this study have been followed up for a year and outcome is being assessed in terms of a second general health questionnaire, consulting and referral patterns and prescribing of psychotropic drugs.

Clarification of the outcome of these disorders in patients seen by general practitioners would provide a better definition of what constitutes a psychiatric 'case' than psychiatric interview alone.

REFERENCES

1. Shepherd, M. and Clare, A. *Psychiatric illness in general practice.* 2nd edition. Oxford: Medical Publications, 1981.
2. Goldberg, D.P. and Blackwell, B. Psychiatric illness in general practice. A detailed study using a new method of case identification. *Br Med J* 1970, 1: 439–443.
3. Marks, J., Goldberg, D.P. and Hillier, V.E. Determinants of the ability of general practitioners to detect psychiatric illness. *Psychol Med* 1979, 9: 337–353.
4. Goldberg, D.P. and Huxley, P. *Mental illness in the community: the pathway to psychiatric care.* London: Tavistock, 1980.
5. Goldberg, D.P., Steele, J.J., Johnson, A. *et al.* Ability of primary care physicians to make accurate rating of psychiatric symptoms. *Arch Gen Psychiatry* 1982, 39: 829–833.
6. Skuse, D. and Williams, P. Screening for psychiatric disorder in general practice. *Psychol Med* 1984, 14: 365–377.
7. Goldberg, D.P., Cooper, B., Eastwood, M.R. *et al.* A standarized psychiatric interview for use in community surveys. *Br J Prev Soc Med* 1970, 24: 18–23.
8. Goldberg, D.P. *The detection of psychiatric illness by questionnaire.* London: Oxford University Press, 1972.
9. Sims, A.C.P. and Salmons, P.H. Severity of symptoms of psychiatric outpatients: use of the general health questionnaire in hospital and general practice patients. *Psychol Med* 1975, 5: 62–66.
10. Tarnopolsky, A., Hand, D.J., McLean, E.K. *et al.* Validity and uses of a screening questionnaire (GHQ) in the community. *Br J Psychiatry* 1979, 134: 508–515.
11. Overton, G.W. and Wise, T.N. Psychiatric diagnosis in family practice: is the general health questionnaire an effective screening instrument? *South Med J* 1980, 73: 763–764.
12. Goldberg, D. Identifying psychiatric illness among general medical patients. *Br Med J* 1985, 291: 161–162.
13. Institute of Psychiatry. *A manual for use in conjunction with the General Practice Research Unit's standardised psychiatric interview.* 2nd edition. London: Institute of Psychiatry, 1970.
14. World Health Organization. *International classification of diseases.* 9th revision. Geneva: WHO, 1978.
15. Royal College of General Practitioners. *Classification of diseases, problems and procedures 1984.* London: RCGP, 1984.
16. Goldberg, D.P. *Manual of the general health questionnaire.* Windsor: NFER, 1978.
17. Goldberg, D. and Bridges, K. Screening for psychiatric illness in general practice: the general practitioner versus the screening questionnaire. *J R Coll Gen Pract* 1987, 37: 15–18.
18. Goldberg, D., Kay, C. and Thompson, L. Psychiatric morbidity in general practice and the community. *Psychol Med* 1976, 6: 565–569.
19. Hoeper, E.W., Nyez, G.R., Cleary, P.D. *et al.* Estimated prevalence of RDC mental disorder in primary medical care. *Int J Mental Health* 1979, 10: 6–15.

20. Goldberg, D. Use of the general health questionnaire in clinical work. *Br Med J* 1986, **293**: 1188–1189.
21. Nott, P.N. and Cutts, S. Validation of the 30-item GHQ in post-partum women. *Psychol Med* 1982, **12**: 409–413.
22. Goodchild, M.E. and Duncan-Jones, P. Chronicity and the general health questionnaire. *Br J Psychiatry* 1985, **146**: 55–61.
23. Freeling, P., Rao, B.M., Paykel, E.S. *et al*. Unrecognised depression in general practice. *Br Med J* 1985, **290**: 1880–1883.

## 5.1 DISCUSSION

### 5.1.1 Objective

The two aims of case study 5 were: to obtain an estimate of conspicuous and hidden psychiatric morbidity in general practice; and to estimate the value of the general health questionnaire (GHQ) in improving case recognition in a clinical setting. What follows concentrates on the second aim partly since the case study does, and also because the main point of including this study in the book is to provide an example for Chapter 13 on the topics of sensitivity and specificity.

### 5.1.2 Design

A group of patients attending one general practitioner completed the 28-item general health questionnaire before seeing their GP, who completed a rating sheet for each patient without seeing their GHQ scores.

### 5.1.3 Subjects

Random samples of 10% taken from all those attending each consulting session of a GP over a five-month period, so that no patient was included more than once, resulted in a total sample of 234 patients. Of these, seven had missing data, leaving 115 males and 112 females.

### 5.1.4 Outcome measures

- Score on the general health questionnaire.
- Rating sheet score based on: reason for consultation (Goldberg and Blackwell categories); degree of psychiatric

disorder (five-point scale); diagnosis (up to two items) and a list of known chronic conditions (up to four).

### 5.1.5 Data and statistical analysis

Table 5.1 summarizes the GP's assessment of the type of disturbance suffered by the 115 male and the 112 female patients in terms of seven categories. Categories 2 to 6 indicate a psychiatric component in 43 (37%) of men and 60 (54%) of women. The narrower description of 'clinical psychiatric illness' was found to apply to only 25 (22%) of men and 35 (31%) or women, and this information is related to GHQ score in Table 5.2. It is the value of the GHQ score, with a cut-off of 9 in deciding whether the GP would assess a patient as having or not having clinical psychiatric illness, which is of interest in Chapter 13, where this application is discussed in detail.

### 5.1.6 Further points

Although the main objective of introducing this case study is as already stated in section 5.1, namely to provide an example of sensitivity and specificity, the clinical implications provided bonuses because 'the doctor took the opportunity to review his diagnosis when his assessment did not agree with the general health questionnaire score' (false positives and false negatives; see Chapter 13). For example, 11 men and 10 women identified as cases by the questionnaire were missed by the doctor. This is an interesting way of using the ideas of sensitivity and specificity in that it differs from the normal use in which the questionnaire might simply be an attempt to act in place of the 'expert' doctor, who is assumed to be 100% certain of identifying cases and non-cases.

# 6

# A randomized controlled trial of surgery for glue ear

Case study 6 is an article by N.A. Black, C.F.B. Sanderson, A.P. Freeland and M.P. Vessey taken from the *British Medical Journal*, **300** (1990), 1551–6.

## ABSTRACT

*Objective*  To assess the effect of five different surgical treatments for glue ear (secretory otitis media) on improvement in hearing and, assuming one or more treatments to be effective, to identify the appropriate indications for surgery.

*Design*  Randomized controlled trial of children receiving (a) adenoidectomy, bilateral myringotomy, and insertion of a unilateral grommet; (b) adenoidectomy, unilateral myringotomy, and insertion of a unilateral grommet; (c) bilateral myringotomy and insertion of a unilateral grommet; and (d) unilateral myringotomy and insertion of a grommet. Children were followed up at seven weeks, six months, 12 months, and 24 months by symptom history and clinical investigations.

*Setting*  Otolaryngology department in an urban hospital.

*Patients*  149 children aged 4–9 years who were admitted for surgery for glue ear and who had no history of previous operations on tonsils, adenoids, or ears and no evidence of sensorineural deafness. Inadequate follow up information on levels of hearing and on middle ear function was obtained from 22.

*Main outcome measures*  Mean hearing loss (dB) of the

three worst heard frequencies between 250 and 4000 Hz, results of impedance tympanometry, and parental views on their child's progress.

*Results* In the 127 children for whom adequate information was available ears in which a grommet had been inserted performed better in the short term (for at least six months) than those in which no grommet had been inserted, irrespective of any accompanying procedure. Most of the benefit had disappeared by 12 months. Adenoidectomy produced a slight improvement that was not significant, though it was sustained for at least two years. The ears of children who had had an adenoidectomy with myringotomy and grommet insertion, however, continued to improve so that two years after surgery about 50% had abnormal tympanometry compared with 83% of those who had had only myringotomy and grommet insertion, and 93% of the group that had had no treatment. Logistic regression analyses identified preoperative hearing level as the single best predictor of good outcome from surgery. Other variables contributed little additional predictive power.

*Conclusions* If the principal objective of surgery for glue ear is to restore hearing then our study shows that insertion of grommets is the treatment of choice. The addition of an adenoidectomy will increase the likelihood of restoration of normal function of the middle ear but will not improve hearing. When deciding appropriate indications for surgery, a balance has to be made between performing unnecessary operations and failing to treat patients who might benefit from surgical intervention. Preoperative audiometry scores might be the best predictor in helping to make this decision.

## INTRODUCTION

Glue ear, or otitis media with effusion, is the commonest reason for elective surgery in childhood.[1] In England and Wales in 1986 about 73 000 operations were carried out in NHS hospitals (based on hospital activity analyses for Oxford and for East Anglian regional health authorities) and a further 18 000 are estimated to have been performed in independent hospitals (J.P. Nicholl, personal communication). Despite the popularity of these operations considerable uncertainty exists about their efficacy and the appropriate indications for their use. Although the results of 15 randomized controlled trials concerning a total of 1549 children have been published since 1967, few of the studies can easily be compared.[2-16] Variations of case definition, exclusion criteria, case severity, outcome measures, duration of follow up, and method of analysis have all contributed to the difficulty in achieving consensus. A further complication is that a variety of operative procedures in different combinations have been studied: adenoidectomy, myringotomy, and grommet (tympanostomy tube) insertion (Table 6.1).

Despite the difficulties entailed in making detailed comparisons between the trials it is possible to identify some consistent findings. First, myringotomy results in little or no benefit.[14,16] Secondly, myringotomy plus grommet insertion is effective for up to 12 months,[2,3,14] though two studies found that this procedure was not effective.[13,15] Thirdly, adenoidectomy is effective,[2,3,12] though again two studies found that it had no effect.[10,11] Fourthly, grommet insertion and adenoidectomy are equally effective,[2,3] though repeat surgery is needed more often after grommet insertion than after adenoidectomy. Finally, adenoidectomy combined with grommet insertion is no better than adenoidectomy alone[2,3,6,8,9] or grommet insertion alone.[2-5,17]

We had two objectives: to compare the relative effectiveness of the five different treatment strategies identified in Table 6.1 and, assuming one or more treatment strategies to be effective, to identify the appropriate indications for surgery in the management of glue ear.

## METHODS

The parents of all children aged 4–9 years who were admitted to the Radcliffe Infirmary, Oxford, for surgery for bilateral glue ear between 1981 and 1986 were invited to allow their child to take part in the trial. Children who had previously had operations on their tonsils, their adenoids, or their ears and those in whom there was evidence of cleft palate or any sensorineural deafness were excluded. Children were also excluded if surgery for conditions other than glue ear was to be performed, such as adenoidectomy for alleviating gross nasal obstruction. The need for surgery was based on the clinical

**Table 6.1** Comparisons considered in published randomised controlled trials of surgery for glue ear, 1967–89

| Treatment 2 | Treatment 1 | | | | |
| --- | --- | --- | --- | --- | --- |
| | No treatment | Myringotomy | Myringotomy and grommet | Adenoidectomy | Adenoidectomy and myringotomy |
| Adenoidectomy and grommet | Maw and Herod[2] | Gates et al.[3] | Roydhouse[4]<br>Widemar et al.[5]<br>Gates et al.[3] | Lildholt[6]<br>Maw and Herod[2] | Richards et al.[7]<br>Bonding et al.[8]<br>Gates et al.[3] |
| Adenoidectomy and myringotomy | | Fiellau-Nikolajsen et al.[10]<br>Gates et al.[3] | Gates et al.[3] | | |
| Adenoidectomy | Rynnel-Dagoo et al.[11]<br>Bulman et al.[12]<br>Maw and Herod[2] | | Maw and Herod[2] | | |
| Myringotomy and grommet | Brown et al.[13]<br>Maw and Herod[2]<br>Mandel et al.[14]<br>Zielhuis et al.[15] | Mandel et al.[14]<br>Gates et al.[3] | | | |
| Myringotomy | Aichard[16] | | | | |

judgement of the otolaryngologist responsible for the care of each child, regardless of any findings on investigation.

Having obtained parental consent for inclusion in the trial, we randomly divided the children into one of four treatment groups: (a) adenoidectomy and bilateral myringotomy plus insertion of a unilateral grommet (standard Shepherd tympanostomy tube); (b) adenoidectomy plus a unilateral myringotomy and insertion of a grommet; (c) bilateral myringotomy plus insertion of a unilateral grommet; and (d) a unilateral myringotomy and insertion of a grommet (Figure 6.1). Randomization between the right ear and the left ear for grommet insertion was also carried out. Instructions about the treatment allocated were contained in sealed numbered envelopes. The contents of the envelopes were determined with a table of random numbers. The clinicians who had obtained parental consent selected the next available envelope according to numerical sequence.

The minimum number of children that would be needed in the study to allow paired analysis and unpaired analysis to be performed was calculated based on the following assumptions. First, we assumed that there would be a mean preoperative variation in hearing loss between a child's ears of 2 (SD 14.25) dB,[13] and, secondly, that there would be a mean preoperative hearing loss of 32.5 (SD 11.4) dB.[18] Finally, we thought that 10 dB should be the minimum difference in levels of hearing between treatments that might be regarded as clinically important and that the trial should have a 95% chance of detecting such a difference between two treatments at the 5% level

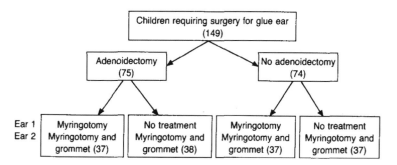

**Figure 6.1** Treatment groups resulting from randomization

of significance. These assumptions implied that about 104 children would be needed for paired analysis – that is, studying the difference in levels of hearing between the two ears in each child – and that about 136 would be needed for unpaired analysis – that is, comparison between treatment groups. We envisaged that about 10% of the children would be lost to follow up before the end of the study, so 149 children were entered into the study.

Information about the child's age, sex, social class (based on the father's occupation, or the mother's when the child was living in a single parent family), and history of symptoms (deafness, otalgia, nasal obstruction, and speech development) was recorded on a preoperative form that was completed by a doctor. In addition, pure tone audiometry (from 250 to 4000 Hz) and impedance tympanometry were carried out. When a myringotomy was performed a record was made of whether the middle ear was dry, contained serous fluid, or contained 'glue'.

Each child was followed up for two years and was reviewed at seven weeks, six months, 12 months, and 24 months. At each visit the following information was obtained: parental views on their child's progress; results of a pure tone audiogram; and results of a tympanogram. The children were not assessed by otoscopy because of the considerable interobserver variation associated with the observations. The audiometricians were blind to the treatment that the children had received. Children who did not attend their follow up appointment were sent another invitation. Attempts to get them to attend were abandoned only when they did not appear at three consecutive appointments.

Parental opinions on their child's treatment were defined as favourable, uncertain, or unfavourable. Parents were also asked to report any adverse side effects of treatment. In line with other trials audiometric performance was based on the mean hearing loss of the three worst heard frequencies.[3,5,6,16] Tympanometry results were classified according to both the shape of the recording and the pressure in the middle ear, and two categories were established: normal (A and C1) or abnormal (B and C2).[19] Tympanometry was not performed on ears with grommets because valid recordings cannot be made when grommets are in place and are patent.

The organizers of the study recognized that after surgery the clinical management of each child remained the responsibility of the otolaryngologist concerned, and therefore any decision to carry out further or repeat surgery was beyond their control. The otolaryngologists were, however, asked to avoid further surgical treatment when possible. Data on repeat surgery were collected and analysed, but the children concerned were no longer followed up.

The statistical analyses consisted of: (a) a comparison of the findings before operation and after operation in the four treatment groups using contingency tables; (b) the proportions of children in each group who had to have repeat surgery, and the findings at reoperation; (c) paired analysis of the audiometric findings for the left ear and the right ear in the same child using *t*-tests on the mean changes in hearing level since surgery; (d) independent comparisons of audiometric findings for the ears of different children after different surgical interventions using *t*-tests on the mean changes in hearing level since surgery; (e) comparison of the proportions of children who had abnormal results on tympanometry and unfavourable parental opinion at follow up; (f) multivariate analysis to link the outcome of grommet insertion to a set of preoperative variables using a range of outcome criteria.

## RESULTS

### Comparability of treatment groups

The children in the four treatment groups were comparable with regard to the stratification criteria of age, social class, and history of glue ear (Table 6.2) and findings on investigation (Table 6.3). The sex ratios differed, but there is no evidence to suggest that this would cause problems with confounding.

### Follow up

Overall 48 (32%) children underwent further surgery for glue ear during the two year follow up period, the proportion varying with the initial treatment group. Children who had undergone an adenoidectomy were less likely to have further surgery (19% versus 45%, $p < 0.01$), but this was not surprising

**Table 6.2** Preoperative characteristics of the children according to treatment group. Values are numbers (percentages)

| Characteristic | Adenoidectomy and bilateral myringotomy plus unilateral grommet (1) (n = 37) | Adenoidectomy plus unilateral myringotomy and grommet (2) (n = 38) | Bilateral myringotomy plus unilateral grommet (3) (n = 37) | Unilateral myringotomy and grommet (4) (n = 37) |
|---|---|---|---|---|
| Social class: | | | | |
| Non-manual | 13 (35) | 16 (42) | 12 (32) | 12 (36) |
| Manual | 18 (49) | 18 (47) | 20 (55) | 21 (56) |
| Other | 6 (16) | 4 (11) | 5 (13) | 3 (8) |
| Pattern of deafness: | | | | |
| Never | 1 (3) | 0 | 2 (5) | 2 (5) |
| Fluctuating | 23 (62) | 21 (55) | 24 (66) | 18 (49) |
| Constant | 13 (35) | 17 (45) | 11 (29) | 17 (46) |
| Duration of deafness (months): | | | | |
| ≤9 | 4 (11) | 7 (19) | 6 (16) | 11 (30) |
| 10–18 | 14 (37) | 13 (34) | 13 (35) | 9 (24) |
| >18 | 19 (51) | 18 (47) | 18 (49) | 17 (46) |

|  | | | | |
|---|---|---|---|---|
| No of episodes of otalgia: | | | | |
| None | 11 (30) | 15 (40) | 12 (32) | 12 (33) |
| 1–3 | 20 (54) | 12 (32) | 16 (42) | 14 (39) |
| ≥4 | 6 (16) | 11 (29) | 9 (26) | 11 (28) |
| Duration of otalgia (months): | | | | |
| <6 | 5 (14) | 2 (5) | 4 (11) | 5 (14) |
| 6–12 | 8 (21) | 6 (17) | 9 (23) | 11 (30) |
| >12 | 24 (64) | 30 (79) | 24 (65) | 21 (57) |
| Nasal symptoms: | | | | |
| None or mild | 11 (30) | 13 (35) | 18 (48) | 16 (44) |
| Moderate or severe | 26 (70) | 25 (65) | 19 (52) | 21 (56) |
| Speech development: | | | | |
| Normal | 30 (81) | 33 (87) | 29 (79) | 33 (90) |
| Abnormal | 7 (19) | 5 (13) | 8 (21) | 4 (10) |
| Mean (SE) age (years) | 6.3 (0.23) | 6.6 (0.23) | 6.1 (0.21) | 6.0 (0.21) |
| Sex (male:female) | 1.06 | 1.11 | 1.92 | 1.79 |

**Table 6.3** Preoperative investigations and operative findings according to treatment group. Hearing level is the mean of the three worst heard frequencies

| Investigation or finding | Treatment group | | | |
|---|---|---|---|---|
| | Adenoidectomy and bilateral myringotomy plus unilateral grommet (1) (n = 37) | Adenoidectomy plus unilateral myringotomy and grommet (2) (n = 38) | Bilateral myringotomy plus unilateral grommet (3) (n = 37) | Unilateral myringotomy and grommet (4) (n = 37) |
| Hearing level (dB): | | | | |
| Left ear | 28.1 | 26.9 | 27.6 | 27.8 |
| Right ear | 29.6 | 29.1 | 29.2 | 27.2 |
| Impedance (No (%) abnormal): | | | | |
| Left ear | 35 (95) | 30 (79) | 31 (84) | 27 (73) |
| Right ear | 35 (95) | 36 (95) | 29 (78) | 29 (78) |
| Middle ear contents (No (%)): | | | | |
| Dry | 11 (30) | 15 (39) | 9 (24) | 16 (44) |
| Serous | 7 (19) | 3 (8) | 4 (10) | 2 (5) |
| Glue | 19 (51) | 20 (53) | 24 (66) | 19 (51) |

as it is usually possible to undergo an adenoidectomy only once. A further 10 (7%) children either did not attend follow up appointments or moved from the area. Most of the loss to follow up occurred more than 12 months after the initial operation: 85% were seen at 12 months but only 61% at 24 months.

### Outcomes

*Audiometry (paired analysis)* Audiometric data were not obtained on every occasion in 22 children and therefore these children were omitted from the analysis. The effect of grommet insertion was assessed in each child by comparing the change in the hearing level (mean of the levels of the three worst heard frequencies) between the ears with and without a grommet. The data were initially analysed without taking into account any loss to follow up. The results are shown in Table 6.4 (raw data). Overall, ears in which a grommet had been inserted performed better in the short term (up to 12 months after surgery) than those in which no grommet had been inserted, irrespective of any accompanying procedures. Losses to follow up occurred for two reasons: repeat surgery and non-attendance at the outpatient clinic. The mean level of hearing of those needing further surgery had deteriorated by about 2 dB during the 12 months since their initial operation compared with an improvement of about 8 dB in those not requiring further intervention. It was possible that those children who had not attended their outpatient appointments had experienced a favourable outcome from surgery. To allow for these potential biases at the 12 and 24 month reviews we modified the raw data by assuming that without repeat surgery the levels of hearing would not have altered from the last recorded level. To test this assumption the analysis was repeated twice, allowing first for a deterioration of 10% in the hearing levels since the last recorded level and then for an improvement of 10%. These variant assumptions made little difference to the results.

*Audiometry (independent comparisons)* Independent (rather than within child) comparisons of changes in mean audiometry scores for the different surgical interventions are shown in Table 6.5. It was apparent that myringotomy had no discernible

**Table 6.4** Within child comparison of change in mean results of audiometry (dB) with time after surgery according to treatment group: raw and modified data. Values are numbers (95% confidence intervals)

| Treatment group comparisons | Time after operation (raw data) | | | | Time after operation (modified data) | |
|---|---|---|---|---|---|---|
| | 7 weeks | 6 months | 12 months | 24 months | 12 months | 24 months |
| Adenoidectomy, myringotomy and grommet versus adenoidectomy and myringotomy | 8.1* (3.0 to 13.3) | 2.8 (−1.9 to 7.4) | −1.0 (−6.1 to 4.0) | 0.7 (−4.9 to 6.4) | −0.9 (−5.5 to 3.8) | 0.2 (−4.9 to 5.3) |
| Adenoidectomy versus adenoidectomy, myringotomy and grommet | 3.3 (−0.5 to 7.1) | 2.8 (−2.2 to 7.8) | 1.9 (−3.6 to 7.4) | 2.2 (−6.0 to 10.3) | 2.3 (−2.8 to 7.4) | 2.1 (−3.8 to 8.1) |
| Myringotomy and grommet versus myringotomy | 12.7* (7.9 to 17.5) | 7.4* (1.4 to 13.4) | 3.7 (−0.4 to 7.8) | 0.9 (−2.7 to 4.6) | 5.5* (0.9 to 10.1) | 3.4 (−1.1 to 8.0) |
| Myringotomy and grommet versus no treatment | 3.4 (−0.9 to 7.6) | 3.5* (0.1 to 6.9) | 1.0 (−2.4 to 4.2) | −2.4 (−8.7 to 3.9) | 2.0 (−1.0 to 5.1) | 0.5 (−3.7 to 4.6) |

*Significant $t$-value ($p < 0.05$).

**Table 6.5** Independent comparisons of changes in mean audiometry scores (dB) with time after surgery: raw and modified data. Values are numbers (95% confidence intervals)

| Treatment group comparisons | Time after operation (raw data) | | | | Time after operation (modified data) | |
|---|---|---|---|---|---|---|
| | 7 weeks | 6 months | 12 months | 24 months | 12 months | 24 months |
| Myringotomy versus no surgery | 1.0 (−4.7 to 6.6) | −0.6 (−7.0 to 5.9) | −1.1 (−8.1 to 5.8) | −2.3 (−9.1 to 4.5) | 1.2 (−5.3 to 7.8) | 0.7 (−5.5 to 7.0) |
| Myringotomy and grommet versus no surgery | 11.7* (5.8 to 17.6) | 8.0* (1.5 to 14.5) | 4.8 (−2.4 to 11.9) | 3.2 (−4.1 to 10.5) | 4.3 (−2.2 to 10.8) | 2.7 (−3.2 to 8.6) |
| Adenoidectomy versus no surgery | 4.5 (−1.3 to 10.4) | 4.3 (−1.4 to 9.9) | 4.3 (−3.1 to 11.6) | 2.4 (−5.7 to 10.5) | 3.2 (−3.5 to 10.0) | 3.5 (−3.2 to 10.3) |
| Myringotomy and grommet versus adenoidectomy | 3.0 (−2.1 to 8.1) | 1.2 (−4.1 to 6.6) | −1.4 (−7.5 to 4.8) | −3.5 (−11.4 to 4.6) | −0.2 (−5.9 to 5.5) | −2.7 (−8.7 to 3.3) |
| Adenoidectomy, myringotomy and grommet versus no surgery | 9.6* (4.3 to 14.8) | 7.6* (2.1 to 13.0) | 5.3 (−1.3 to 11.9) | 5.9 (−1.9 to 13.3) | 4.6 (−1.3 to 10.4) | 5.9 (−0.2 to 12.0) |
| Adenoidectomy, myringotomy and grommet versus adenoidectomy | 6.9* (0.8 to 13.0) | 3.8 (−2.6 to 10.2) | 0.0 (−4.0 to 4.0) | 4.3 (−4.4 to 13.0) | 0.3 (−6.8 to 7.4) | 2.6 (−4.7 to 9.8) |
| Adenoidectomy, myringotomy and grommet versus | 2.0 (−2.3 to 6.4) | 2.1 (−2.6 to 6.8) | 2.4 (−2.7 to 7.6) | 6.9* (0.3 to 13.7) | 1.5 (−3.3 to 6.4) | 5.1 (0.0 to 10.2) |

effect compared with no treatment. In contrast, levels of hearing improved with both myringotomy plus grommet insertion and adenoidectomy. The outcome after the combined operation (adenoidectomy plus myringotomy and grommet insertion) confirmed these findings. There was little difference initially between the outcome of the combined operation and that obtained with myringotomy and grommet insertion alone (Figure 6.2). Sensitivity analysis with different modifications to the raw data (again allowing for a deterioration of 10% in the levels of hearing since the last recorded level and an improvement of 10%) made little impact on the results, and comparisons of absolute values of the levels of hearing on follow up, rather than changes from the preoperative levels, produced similar findings.

*Tympanometry* In addition to the difficulties caused by the fairly high drop out rate during the second year of follow up the results of tympanometry were also affected by the lack of data

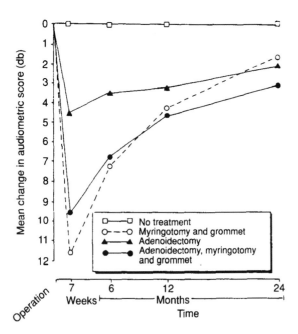

**Figure 6.2** Independent comparisons of changes in mean audiometry scores for several different treatments using modified data

during the first year of follow up for those ears in which a grommet had been inserted (because tympanometry could not be performed as a satisfactory seal cannot be achieved after grommet insertion). Because myringotomy had no effect on the levels of hearing the findings on tympanometry were analysed according to four groups (Figure 6.3). Because of the pre-operative differences in the proportions of abnormal readings changes in proportions were used in the analysis. During the second year the ears of children who had had an adenoidectomy continued to improve so that two years after surgery about half of them had abnormal tympanograms compared with 83% of those who had had a myringotomy plus grommet insertion, and 93% of those who had had either a myringotomy or no treatment.

*Parental opinion* It is difficult to assess the state of each of their child's ears separately. Parental opinion could therefore be used as an outcome measure only in relation to the four treatment groups. The parents of children who had had an adenoidectomy were more satisfied than those whose children

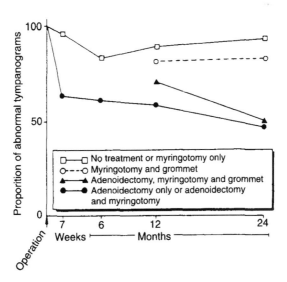

**Figure 6.3** Proportion of ears with abnormal impedance (B and C2) preoperatively that remained abnormal postoperatively for different treatment groups using modified data. No data were available for ears in which a grommet was in place

had not (Figure 6.4). This difference persisted throughout the two years of follow up so that by the end of the second year about half of the children who had not had an adenoidectomy were thought to be satisfactory compared with around 60–70% of those who had.

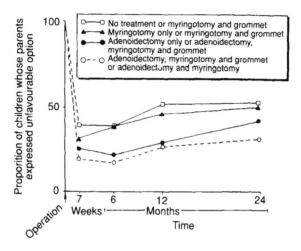

**Figure 6.4** Proportion of children whose parents thought that their child's condition was unfavourable or uncertain at various times postoperatively according to treatment group

### Indications for surgery

Logistic regression analyses were carried out to establish the appropriate indications for inserting grommets with or without adenoidectomy. The predictive power of a wide range of variables was considered: patient characteristics (age, sex, social class); symptoms (deafness, otalgia, nasal obstruction, speech); findings on investigation (hearing level, impedance); and findings at operation (middle ear contents). With the outcome criterion being defined as a relative improvement in hearing level of 10 dB after 12 months in the ear that had a grommet compared with the ear that did not, the data were examined for a subset of variables that had some predictive power. The most useful were the preoperative hearing level and the contents of the middle ear. Other variables contributed little additional predictive power. As the purpose of this analysis was to provide a basis for decisions about whether to operate, further analyses omitted the contents of the middle ear as this information may be reliably obtained only during surgery.

The accuracy of using preoperative audiometry scores as the sole predictor of outcome was tested using various different mean (for left and right ears together) preoperative hearing levels as indicative of surgery and two levels of improvement (5 dB and 10 dB), at six and 12 months after the operation, as indicative of a satisfactory outcome (Table 6.6). At six months the proportion of children who had an improvement of 10 dB or more was 38% among those whose preoperative hearing loss was 25 dB or more (95% confidence interval 27% to 50%). At 12 months this had dropped to 29% (95% confidence interval 19% to 40%). The corresponding figures among those whose preoperative hearing loss was less than 25 dB was 8% at both six months and 12 months (95% confidence interval 3% to 20%).

## DISCUSSION

This trial was designed to assess the effectiveness of surgery for glue ear rather than its efficacy. As such, no attempt was made to alter existing clinical practice – for example, by insisting that highly experienced senior surgeons assessed the children and performed the operations. Most of the surgery was performed by senior house officers and, with a steady turnover of medical staff, around 15 doctors of different grades were involved in the preoperative and postoperative care and assessment of patients. Recruitment of children took considerably longer than expected. This was due to failure by junior medical staff to attempt to recruit patients rather than a poor response rate. A comparison of the characteristics of the children included in the trial with those of the population of children undergoing surgery[20] suggested that those included were representative and that no selection bias had operated. We believe that the clinical management the children experienced was fairly typical of otolaryngological practice in England and Wales in the 1980s. The results obtained are therefore likely to reflect the effectiveness of current practice.

**Table 6.6** Comparisons of power of different preoperative levels of hearing to predict relative improvements mean audiometric score of at least 5 dB and at least 10 dB six and 12 months after surgery. Predictor refers mean preoperative level of hearing taken to indicate operation

| Predictor (dB) | Operation indicated | | Operation not indicated | | % Of whole group denied benefit from surgery (95% confidence interval) |
|---|---|---|---|---|---|
| | No | % with good outcome (95% confidence interval) | No | % with good outcome (95% confidence interval) | |
| *Improvement ≥10 dB six months postoperatively* | | | | | |
| 30 | 53 | 45 (32 to 59) | 74 | 14 (7 to 24) | 8 (4 to 14) |
| 25 | 79 | 38 (27 to 50) | 48 | 8 (3 to 21) | 3 (1 to 8) |
| 20 | 96 | 33 (25 to 44) | 31 | 6 (1 to 23) | 2 (0 to 6) |
| 15 | 117 | 28 (20 to 37) | 10 | 10 (0 to 46) | 1 (0 to 5) |
| *Improvement ≥10 dB 12 months postoperatively* | | | | | |
| 30 | 53 | 36 (23 to 50) | 76 | 11 (5 to 20) | 6 (3 to 12) |
| 25 | 80 | 29 (19 to 40) | 49 | 8 (3 to 20) | 3 (1 to 8) |
| 20 | 100 | 25 (17 to 35) | 29 | 7 (1 to 24) | 2 (0 to 6) |
| 15 | 121 | 22 (15 to 31) | 8 | 0 (0 to 40) | 0 (0 to 4) |
| *Improvement ≥5 dB six months postoperatively* | | | | | |
| 30 | 53 | 57 (42 to 70) | 74 | 34 (23 to 46) | 20 (13 to 28) |

The only important methodological problem experienced was the higher than predicted number of children whom we were unable to follow up for two years. The principal reason for this was the clinicians' (and occasionally the parents') dissatisfaction with a child's progress, which they believed warranted further surgical intervention. To cope with this problem the data were modified in the way we described. The results obtained with sensitivity analysis were robust to the various assumptions we made about those lost to follow up. Nevertheless, it is necessary to bear this adjustment in mind, particularly when considering data that related to the two year follow up.

It was clear that myringotomy plus grommet insertion produced a significant improvement in hearing which lasted for six to 12 months. Adenoidectomy resulted in only a modest improvement in hearing, though there was some evidence to suggest this was more long lasting than that obtained from the insertion of grommets. This view was supported by the finding that normal function of the middle ear (measured by impedance tympanometry) was restored in about half the children who underwent an adenoidectomy compared with only about 20% of children after myringotomy plus grommet insertion. If, however, the primary objective of surgery for glue ear is to restore hearing then this apparent advantage of adenoidectomy is irrelevant. To achieve a rapid and significant improvement in hearing myringotomy plus grommet insertion is the treatment of choice. The addition of an adenoidectomy produces little additional benefit. In this respect the results of this trial are consistent with those of several other studies.[2-6,8,9] Considering the operative risks and the greater economic and social costs of adenoidectomy compared with myringotomy plus grommet insertion, our results offer little justification for continuing to use adenoidectomy in the routine treatment of glue ear. The finding that the proportion of parents who expressed satisfaction with the treatment that their child had received was higher among those whose children had had an adenoidectomy than in those whose children had not, might be explained by the first group's knowledge that everything that might have been done had been done.

The need for clinicians to identify those children who would benefit from surgery is clear. Unfortunately, none of the 15 published randomized, controlled trials has considered the issue quantitatively. Our study has, however, investigated the sensitivity and specificity of preoperative findings in predicting the outcome of surgery. Despite the uncertainties surrounding the level of objectivity of audiometry this single measure appears to be a useful predictor of outcome. The use of preoperative hearing level both for ears that had grommets inserted and those that had not should have inhibited the effects of regression towards the mean.

Interpretation of the preoperative audiometry score as a predictor of outcome of surgery depends on the definition of a satisfactory outcome in terms of improvement in hearing and on attitudes to unnecessary operations on the one hand and to missed cases (children who might have benefited from surgery but who were not treated) on the other. The confidence intervals from this study were wide, but the implications for current practice are potentially dramatic. For example, if satisfactory outcome is defined as an improvement of 10 dB six months after surgery, and if a strategy of operating only on children with a hearing loss of 25 dB and above is adopted, then only 79 of the 127 children with complete data in this trial would have been operated on, of whom it might be expected that 30 would have benefited and 49 would not. Four children, however, who might have benefited would have been missed. Alternatively, setting the operation threshold at 20 dB would have resulted in 96 operations being performed, with 32 children expected to benefit, and two potential beneficiaries being missed. If the children in this trial were representative of children operated on for glue ear in England and Wales in 1986 then the adoption of a policy of only operating when the preoperative hearing loss is at least 25 dB would have had the following implications. First, the total number of operations would have been reduced from 91 000 to about 57 000, of which 21 000 would have achieved a satisfactory improvement of 10 dB or more. Secondly, however, nearly 3000 of the 34 000 children who would have been regarded as inappropriate for surgery under this policy would have been denied such an improvement.

These figures give only an indication of the scale of the problem. As in any trial, the sample used might not have been representative of the population of children undergoing sugery for glue ear and the effectiveness of the surgeons concerned might not have been typical. Also predictors that have been derived from one set of patients will generally not perform as well when used with another set, and greater precision is required. It will be necessary to test the predictors on other samples of children to confirm our results.

Finally, it is important to recognize that, as with most trials of surgery for glue ear, the effectiveness of the operations was assessed in terms of improvement in hearing. No attempt was made to determine any possible longer term effects – namely, improvements in language skills or in educational achievements.

## REFERENCES

1. Black, N.A. Surgery for glue ear – a modern epidemic. *Lancet* 1984, i: 835–837.
2. Maw, A.R. and Herod F. Otoscopic, impedance, and audiometric findings in glue ear treated by adenoidectomy and tonsillectomy. A prospective randomised study. *Lancet* 1986, i: 1399–1402.
3. Gates, G.A., Avery, C.A., Prihoda, T.J. *et al.* Effectiveness of adnoidectomy and tympanostomy tubes in the treatment of chronic otitis media with effusion. *N Engl J Med* 1987; **317**: 1444–1451.
4. Roydhouse, N. Adenoidectomy for otitis media with mucoid effusion. *Ann Otol Rhinol Laryngol* 1980, **89**(suppl 68): 312–315.
5. Widemar, L., Svensson, C., Rynnel-Dagoo, B. *et al.* The effect of adenoidectomy on secretory otitis media: a two year controlled prospective study. *Clin Otolaryngol* 1985, **10**: 345–350.
6. Lildholdt, T. Unilateral grommet insertion and adenoidectomy in bilateral secretory otitis media: preliminary report of the results in 91 children. *Clin Otolaryngol* 1979, **4**: 87–93.
7. Richards, S.H., Kilby, D., Shaw, J.D. *et al.* Grommets and glue ears: a clinical trial. *J Laryngol Otol* 1971, **85**: 17–22.
8. Bonding, P., Tos, M. and Poulsen, G. Unilateral insertion of grommets in bilateral secretory otitis media. *Acta Otolaryngol* 1982, Suppl **386**: 161–162.
9. To, S.S., Pahor, A.L. and Robin, P.E. A prospective trial of unilateral grommets for bilateral secretory otitis media in children. *Clin Otolaryngol* 1984, **9**: 115–117.
10. Fiellau-Nikolajsen, M., Hojslet, P.E. and Felding J.U. Adeno-idectomy for eustachian tube dysfunction: long term results from a randomised controlled trial. *Acta Otolaryngol* 1982, Suppl **386**: 129–31.
11. Rynnel-Dagoo, B., Ahlborn, A. and Schiratzki H. Effects of adenoidectomy. A controlled two year follow up. *Ann Otol Rhinol Laryngol* 1978, **87**: 272–278.

12. Bulman, C.H., Brook S.J. and Berry, M.G. A prospective randomised trial of adenoidectomy versus grommet insertion in the treatment of glue ear. *Clin Otolaryngol* 1984, **9**: 67–75.
13. Brown, M.J.K.M., Richards, S.H. and Ambegaokar, A.G. Grommets and glue ear: a five year follow up of a controlled trial. *J R Soc Med* 1978, **71**: 353–356.
14. Mandel, E.M., Bluestone, C.D., Paradise, J.L. *et al.* Efficacy of myringotomy with and without tympanostomy tube insertion in the treatment of chronic otitis media with effusion in infants and children: results for the first year of a randomised clinical trial. In: Lim, D.J., Bluestone, C.D., Klein, J.O. *et al* eds. *Recent advances in otitis media with effusion.* Philadelpia: Decker, 1984: 308–312.
15. Zielhuis, G.A., Rach, G.H. and Van Den Broek, P. Screening for otitis media with effusion in preschool children. *Lancet* 1989, i: 311–314.
16. Archard, J.C. The place of myringotomy in the management of secretory otitis media in children. *J Laryngol Otol* 1967, **81**: 309–315.
17. Black, N.A., Crowther, J. and Freeland, A. The effectiveness of adenoidectomy in the treatment of glue ear: a randomised controlled trial. *Clin Otolaryngol* 1986, **11**: 149–155.
18. Tos, M. and Poulsen G. Secretory otitis media. Late results of treatment with grommets. *Archives of Otolaryngology* 1976, **102**: 672–675.
19 Paradise, J.L., Smith, C.G. and Bluestone, C.B. Tympanometric detection of middle ear effusion in infants and young children. *Prediatrics* 1976, **58**: 198–210.
20. Black, N.A. The aetiology of glue ear – a case control study. *Int J of Pediatr Otorhinolaryngol* 1985, **9**: 121–133.

## 6.1 DISCUSSION

### 6.1.1 Objective

To assess the effect of different surgical treatments for glue ear.

### 6.1.2 Design

A randomized controlled trial (RCT) of children allocated at random to one of four groups (Table 6.7).

The study size – that is, the number of subjects required for the trial – was determined by considering the preoperative variation in hearing loss between a child's ears, the preoperative hearing loss, the minimum difference in levels of hearing between treatments that might be regarded as clinically im-

**Table 6.7** Allocation of subjects to the four groups

|  | Group | | | |
|---|---|---|---|---|
|  | *1* | *2* | *3* | *4* |
| Treatment | a | a | not a | not a |
| Ear 1* | m | not m<br>not g | m | not m<br>not g |
| Ear 2* | m + g | m + g | m + g | m + g |

a = adenoidectomy; m = myringotomy; g = grommet inserted.
*The numbering of each child's ears was done randomly.

portant, and the requirement that the trial should have a 95% chance of detecting this minimum difference between two treatments at the 5% level of significance. The inclusion of these considerations in determining the study size in case study 6 is the main reason for including this case study in this book. The details of how the calculations of study size were probably made are given in Chapter 14.

### 6.1.3 Subjects

The calculations referred to above resulted in a total study size of 149 children, who were allocated randomly to the four treatment groups, giving 37, 38, 37, 37, respectively.

### 6.1.4 Outcome measures

- Each child's age, sex, social class, history of symptoms, pure tone audiometry and impedance tympanometry.
- Follow-up at 7 weeks, 6 months, 12 months and 24 months recorded parental views on their child's progress, results of a pure tone audiogram, and results of a tympanogram.

### 6.1.5 Data and statistical analysis

Apart from the preoperative data given in Tables 6.2 and 6.3, it is Tables 6.4 and 6.5 which show comparisons between pairs of treatments. Table 6.4 gives details of 'paired' i.e. within-child comparisons of change in mean results of audiometry, using the methods of section 8.7, in which one of the paired $t$-tests reported in this case study is given in detail; Table 6.5 gives details of 'unpaired', that is, independent or 'between-children' comparisons of changes in mean audiometry scores, using the methods of section 8.8. (The topic of logistic regression, mentioned in case study 6 in the section on 'Indications for Surgery', is beyond the scope of this book.)

### 6.1.6 Further points

The main conclusions of this case study concern the effectiveness of different types of surgery. These conclusions are based on the paired and unpaired confidence intervals and hypothesis tests reported in Tables 6.4 and 6.5. Notice that those comparisons with significant $t$-values ($p < 0.05$) are those for which the corresponding 95% confidence interval does *not* contain the value 0. The connection between the confidence interval approach and the hypothesis test approach to statistical inference is discussed in section 8.6.

# Part Two

---

# Statistical Methods

*Go with me, Lord Goschen said, into the study of statistics, and I will make you all enthusiasts in statistics.*

# 7

# Data, tables, graphs, summary statistics and probability

*When she [FN] had finished the statistical section of her Report, she sent the proofs with her illustrative diagrams for Dr Farr's revision. He [the statistician in the Registrar-General's office] found nothing to alter. 'This* speech*', he wrote, 'is the best that ever was written on Diagrams or on the Army. I can only express my Opinion briefly that Demosthenes himself with the facts before him could not have written or thundered better'. He especially commended her diagrams for the clearness with which they explained themselves. She was something of a pioneer in the graphic method of statistical presentation.*

## 7.1 DATA

Statistical data collected from individual subjects or patients are either **numerical** or **non-numerical**. Age is a numerical variable, while gender is a non-numerical or **categorical** variable. The distinction is not always clear-cut. For example, smoking habits can be expressed in terms either of 'the number of cigarettes smoked per day', a numerical variable, or in terms of the categorical variable 'whether subject smokes', with categories 'no' and 'yes'.

The type of data we collect for individual subjects determines to some extent how we summarize data for a group of subjects. So we may wish to find the mean age of all subjects in a control group, while we may wish to find the percentage of females in the same control group. Section 7.4 deals with summary statistics in more detail.

### 7.1.1 Examples from the case studies

#### *Case study 1*

From Table 1.1, it is clear that the following variables were collected for each subject: age, weight, height. All these are numerical variables. Data were also collected on class (since social class is determined by type of employment, it is essentially non-numerical), whether hypertension is diagnosed, whether the subject currently smoked, whether the subject has ever smoked, gender, group. All these are non-numerical variables. Table 1.5 also gives information for a further five numerical variables.

#### *Case study 2*

In Tables 2.1 and 2.2 we find a number of non-numerical variables:

| *Variable* | *Categories* |
|---|---|
| Group | Intervention, control |
| Attenders | Yes, No (intervention group only) |
| Smoking at one of four stated times | Yes, No |
| CO monitoring | Yes, No (attenders only) |

#### *Case study 3*

The numerical variables are: general health questionnaire score (Table 3.1); and the number of consultations per year. The non-numerical variables are: gender (Tables 3.1–3.4); diagnostic label (Table 3.2); and psychotropic drugs (Table 3.3).

#### *Case study 4*

There is a complication here since the individual 'subjects' in this paper refer to the 98 family practitioner committees. The variables used therefore refer to the 98 groups of patients – *not* the individual patients. Some of these variables are: standardized mortality ratio, Jarman score, number of prescriptions per patient, cost of prescriptions per patient. All of these are numerical variables.

I leave it as an exercise for the reader to find other examples in case studies 5 and 6.

## 7.2 TABLES AND GRAPHS

When the results of a medical study are presented in a paper, the focus is usually on groups of subjects rather than on individuals. Tables of summary statistics classified by row and column headings are very popular in medical journals (there are 25 such tables in the case studies), while graphs (or 'figures') are less popular (only seven in all, four of which occurred in one case study). The reason for the dearth of pictorial presentation is presumably the cost in terms of time and money. This is a pity because graphs can be very informative, although their interpretation may, to some extent, be subjective.

### 7.2.1 Examples from the case studies

There are no graphs in case studies 1, 2 or 3.

### *Case study 4*

Figure 4.1(a) immediately conveys something of the relationship between the number of prescriptions per patient and standardized mortality rate for each of 98 family practitioner committees. Notice:

- A tendency for high SMRs to be associated with high prescription rates.
- However, there is a lot of scatter ('noise') in the relationship).
- How the line drawn through the points influences your conclusions about the connection between the variables. Do you wonder how the position of the line was determined? (See Chapter 12 of this book in due course!).
- In the test that $\bar{R}^2 = 0.46$, $p < 0.001$ is quoted in relation

to this graph. What does this mean? (Again Chapter 12 will explain.)

### Case study 5

Here the only graph shows the relationships between sensitivity and specificity to general health questionnaire score, and implies an inverse relationship (Chapter 13). Again this is a very effective way of presenting data.

### Case study 6

Figure 6.1 is effective here in explaining the design of the randomized control trial used in terms of treatments and numbers of patients per group. Figures 6.2–6.4 all represent plots of various variables against time for different treatments. For example, Figure 6.2 shows mean change in audiometric score for four treatments. One can gain the impression that myringotomy and grommet together are effective, but the additional effect of adenoidectomy is negligible. This impression is, of course, subjective; more objective analysis is shown in Table 6.5 where a large number of $t$-tests are carried out (Chapter 8).

### 7.3 SUMMARY STATISTICS FOR NUMERICAL VARIABLES

Summary statistics in medicine are often numbers which combine data from groups of subjects. The particular statistics used depend on the type of data collected (numerical or non-numerical) and the purpose for which the statistics have been calculated.

For a numerical variable such as age it may be appropriate to calculate: the mean age (years) and/or the standard deviation of age (years). The purpose of the mean is to provide one value which represents the centre of the data. The standard deviation is a measure of variation of the data about the mean. These (the mean and standard deviation) will be appropriate if the distribution of the individual values of the variable is reasonably symmetrical when plotted on a dotplot, as in Figure 7.1 (fictitious data for a random sample of 50 subjects from some

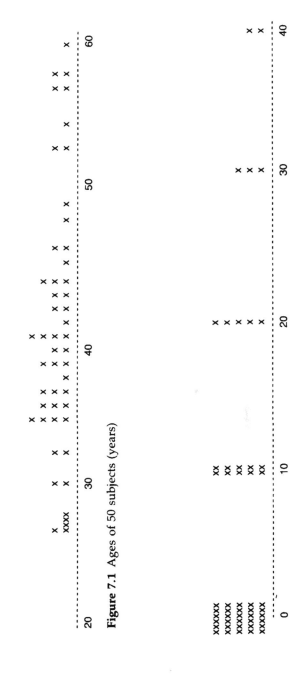

**Figure 7.1** Ages of 50 subjects (years)

**Figure 7.2** Cigarettes per day smoked by 50 subjects

population). The mean appears to be about 40 years, with roughly equal numbers of subjects below and above 40.

However, the distribution of the number of cigarettes smoked per day shown in Figure 7.2 is not at all symmetrical, and to say 'the mean number of cigarettes smoked per day' = 7.4 is not very helpful.

Sometimes the median is suggested instead of the mean for very skew data. The calculation of the median requires that the data are put in order of magnitude and then the median value is the 'middle' one. However, this gives a median of 0 for the data in Figure 7.2 (since both the 25th and 26th values are 0).

Perhaps the best we can do here is to say that:

> 60% of subjects do not smoke, while for smokers the mean number of cigarettes per day is 18.5.

The moral is to plot the distributions of all numerical variables, but how often do we see any hint of this in a published paper? Never!

Assuming the mean is the correct measure of central tendency, how do we calculate it? The answer is: by using a calculator (or computer) and, after entering the data, pressing the $\bar{x}$ button, where

$$\bar{x} = \frac{\Sigma x}{n}$$

meaning 'sum the $n$ data values and divide by $n$'. We call $n$ the 'sample size', and $\bar{x}$ is called the 'sample mean'.

How do we calculate the standard deviation, and what does it mean when we calculate it? The answer is: by using a calculator (or computer) and, after entering the data, pressing the $s$ button (some calculators call this $\sigma_{n-1}$ instead of $s$), where

$$s = \sqrt{\frac{\Sigma(x - \bar{x})^2}{n - 1}}$$

The method implies that we subtract $\bar{x}$, the sample mean, from each value in turn, square these differences, add them all up, divide by $n - 1$ and take the square root. Three reasonable questions are:

- Why is it so complicated, why not *use* $\Sigma(x - \bar{x})$ as a measure of variation about the mean? Unfortunately, this will always give the same answer, namely 0.

*Example 7.1*    117

- Why divide by $n - 1$ and not $n$? The answer is beyond the scope of this book, but see Altman (1991: 34), Armitage and Berry (1987: 36), Bland (1987: 66) or Rees (1989: 33) – these references are listed in Appendix A.
- Why take the square root? The answer is 'to make the units of standard deviation the same as the units of the variable in question'. So if the variable is age in years, the mean will be in terms of years and so will the standard deviation (see Example 7.1 below).

In some analyses we will wish to deal with $s^2$, the square of the standard deviation, which is called the **variance** (see Chapter 9, for example).

### EXAMPLE 7.1

Calculate the mean and standard deviation of the ages of the 50 subjects in Figure 7.1. Suppose the ages are:

| 26 | 26 | 27 | 28 | 29 | 30 | 30 | 32 | 32 | 34 |
|----|----|----|----|----|----|----|----|----|----|
| 34 | 34 | 34 | 35 | 35 | 35 | 36 | 36 | 36 | 37 |
| 38 | 38 | 38 | 39 | 39 | 40 | 40 | 40 | 40 | 41 |
| 42 | 42 | 43 | 43 | 44 | 44 | 44 | 45 | 46 | 46 |
| 48 | 49 | 53 | 53 | 55 | 57 | 57 | 58 | 58 | 60 |

Entering these numbers into a calculator, we obtain

$\bar{x} = 40.52$ years
$s = 8.96$ years
$n = 50$

We can understand why 40.52 is the mean age by reference to Figure 7.1 since it is in the middle of a relatively symmetrical distribution. But, what does $s = 8.96$ years tell us? It is simply a measure of variation about the mean: the larger the value of the standard deviation the larger the variation about the mean. When the distribution of the data has a particular bell shape, called the normal distribution (section 7.5), we should find the following results apply, at least approximately:

- About two-thirds (68%) of the values should lie within one standard deviation of the mean.
- Nearly all (theoretically 95%) of the values should lie within two standard deviations of the mean.

Do these results apply to our example?

$$\bar{x} - s = 31.6, \qquad \bar{x} + s = 49.5$$
$$\bar{x} - 2s = 22.6, \qquad \bar{x} + 2s = 58.4$$

Since 35 out of 50 (70%) lie in the range 31.6 to 49.5, while 49 out of 50 (98%) lie in the range 22.6 to 58.4, it would be fair to conclude the distribution of age is at least approximately 'normal'. Clearly the data in Figure 7.2 are not 'normal' so we would not expect agreement with theory in this case. The reader is invited to check this by calculation. □

So far in this section we have considered summary statistics for numerical variables, and the terms mean and standard deviation have been introduced. We find the term 'mean' in abundance in medical papers, but the term 'standard deviation' is not mentioned in any of the six case studies. Instead, we find the term 'standard error' mentioned, particularly in case studies 1 and 3. Why is this?

In fact, the standard error in these examples is simply the standard deviation divided by the square root of the sample size. In symbols:

$$se = \frac{s}{\sqrt{n}}$$

Here we should use the term 'standard error of the mean' and not just 'standard error'! In fact the standard error of the mean is the standard deviation of the sample mean. Theory (called the central limit theorem) indicates that the sample mean (of samples taken from a population) will be normally distributed. Hence we can say that 95% of sample means will lie with two standard errors of the population mean as long as $n$, the sample size, is large (greater than 30, say). This statement can be turned around to state that, if we take a large sample from a population, and calculate the sample mean ($\bar{x}$) and the standard error ($se = s/\sqrt{n}$), then there is a 95% chance that the population mean will lie between ($\bar{x} - 2se$) and ($\bar{x} + 2se$). Note the following observations:

- These two values are called 95% confidence limits (Chapter 8).

- To be strictly accurate, 1.96 should be used instead of 2.
- The formula above only applies if $n$ is large (section 8.4)
- The standard error is in the same units as the standard deviation, that is, the same as the units of the variable in question.

### EXAMPLE (7.1 revisited)

The standard error of the mean is $se = 8.96/\sqrt{50} = 1.27$ years. Thus $\bar{x} \pm 2se$ gives $40.52 \pm 2 \times 1.27$ *or* 38.0 to 43.1 years. The values of 38.0 and 43.1 are the 95% confidence limits for the mean age of the population (from which the random sample of 50 subjects was drawn). Note that $n = 50$, a largish sample. □

### 7.3.1 Examples from the case studies

*Case study 1*

Table 1.1 gives means and standard errors for three variables, namely age, weight and height, for each of four groups of subjects. Table 1.4 gives means and standard errors for total dietary fibre at four time periods, for each of four groups of subjects.

*Case study 3*

Table 3.1 gives means and standard errors for total GHQ score, and also for four sub-total GHQ scores.

### 7.4 SUMMARY STATISTICS FOR NON-NUMERICAL VARIABLES

When we collect data for a non-numerical (categorical) variable for a number of subjects, we will probably wish to know how many subjects fall into each of the possible categories. Using the data from Figure 7.2 and the categories 'non-smoker' and 'smoker', we could draw up the following table of smoking habits for 50 subjects.

| Number of subjects (%) | |
|---|---|
| Non-smokers | Smokers |
| 30 (60%) | 20 (40%) |

The percentages are obtained by multiplying proportions by 100, so

$$\frac{30}{50} \times 100 = 60\% \quad \text{and} \quad \frac{20}{50} \times 100 = 40\%$$

When a variable has more than two categories, there will be more than two columns. When we wish to summarize data for two categorical variables we can form a two-way (or 'contingency') table using the rows for the categories of one variable and the columns for the other variable.

### 7.4.1 Examples from the case studies

*Case study 3*

Table 7.1 is a simplified version of Table 3.3. Notice that the row categories are mutually exclusive (that is, non-overlapping), as are the column categories. Percentages can be calculated in three ways:

- As a proportion of the grand total.
- As a proportion of the row table.
- As a proportion of the column total.

**Table 7.1** Drugs prescribed to the patients receiving a diagnosis of psychiatric illness in the follow-up year

| | Number of patients | |
|---|---|---|
| | Men | Women |
| Drug 1 | 9 | 12 |
| Drug 2 | 4 | 17 |
| Both drugs | 4 | 6 |
| No drugs | 3 | 0 |

The last of these is used in case study 3.

Notice that Tables 3.2 and 3.4 are similarly constructed to Table 3.3, but it is not at all clear that the categories of Table 3.5 are mutually exclusive. In fact, Table 3.5 is really four tables in one (one per row).

### Case study 5

In Table 5.1, you can tell that the seven row categories are mutually exclusive, because the frequencies add up to a total, and the percentages add up to 100 (separately for men and women).

### Case study 6

Table 6.2 is really seven tables in one. For example a 3 × 4 table could be formed for the variables social class (rows) and treatment group (columns).

### Case study 2

In more advanced tables, such as Tables 2.1 and 2.2, percentages are quoted for the sample of subjects who took part in the study, along with 95% confidence limits for the percentage in the population who are estimated to fall into each of the cross-categories. Confidence limits for this type of data are discussed in Chapter 10.

### 7.5 PROBABILITY

In the next chapter and in much of the rest of this book we will be discussing inferences about a population based on sample data. Since we usually cannot measure the whole of a population (because it would cost too much and take too long), we rely on sample data. It follows that any statements we make about the population are subject to some uncertainty. A way of measuring uncertainty is in terms of probability, and this measure lies between 0 and 1.

We can often estimate the probability of an event by calculating how often the event happens as a proportion of the number of times it could happen. This 'relative frequency' is a reasonable estimate if we collect enough relevant data.

For example, of 1000 live births, 497 girls are born. We estimate the probability that a girl is born by dividing the number of girls born in our sample by the total number of live births in the sample. Thus

$P$ (girl) $= 0.497$

Of 1000 deaths, 245 died of heart problems and 2 were killed by lightning. Thus

$$P \text{ (death due to heart problems)} = \frac{245}{1000} = 0.245$$

$$P \text{ (death due to lightning)} = \frac{2}{1000} = 0.002$$

Of 100 000 male adults, all are less than 7 feet tall. Thus

$$P \text{ (height less than 7 feet)} = \frac{100\,000}{100\,000} = 1$$

Notice that probabilities can be translated into percentages by simply multiplying by 100. For example, 49.7% of births result in girls, and so on.

Suppose we collect the heights of a very large sample of male adults, so large we can consider it as a 'population of heights'. Suppose we also find three other facts:

- The population mean height $\mu$ is 170 cm.
- The standard deviation of height $\sigma$ is 10 cm.
- The distribution of height* conformed to the normal distribution.

Notice the use of the Greek letter $\mu$ for *population* mean, while we used $\bar{x}$ to represent the *sample* mean. Similarly, we use the Greek letter $\sigma$ for *population* standard deviation, while we use $s$ to represent the *sample* standard deviation.

---

* Certain anatomical variables have been found by many reseachers to be 'normally distributed'. The height of a well-defined human population is one such variable.

*Example 7.2* 123

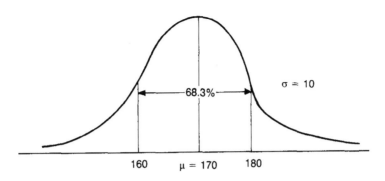

**Figure 7.3** A normal distribution with $\mu = 170\,$cm, $\sigma = 10\,$cm

Then we can draw a graph such as Figure 7.3. This figure measures probability in terms of area under the curve. So the total area is 1, while we can quote three standard results (which apply to *any* normal distribution) and how they apply to the example shown in Figure 7.3. (two of these results were mentioned in section 7.3)

First, 68.3% of a normal distribution lies within one standard deviation of the mean. Thus 68.3% of male adults are between 160 cm (170 − 10) and 180 cm (170 + 10) tall.

Second, 95% of a normal distribution lies within 1.96 standard deviation of the mean. Thus 95% of male adults are between 150.4 cm and 189.6 cm tall.

Third, 99.8% of a normal distribution lies within 3.09 standard deviations of the mean. Thus 99.8% of male adults are between 139.1 cm and 200.9 cm tall.

Other results can be worked out using Table B.1(a) (Appendix B) and noting the symmetry of the normal distribution.

### EXAMPLE 7.2

For a normal distribution with a mean $\mu = 170\,$cm, standard deviation $\sigma = 10\,$cm, what percentage will have heights:

(i) less than 150 cm?
(ii) greater than 200 cm?
(iii) between 150 and 200 cm?

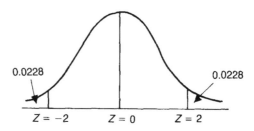

**Figure 7.4** A standardized normal distribution

To tackle (i), let $x = 150$. Calculate

$$z = \frac{x - \mu^*}{\sigma} = \frac{150 - 170}{10} = -2$$

The number 2 implies we are 2 standard deviations from the mean, the negative sign means we are below the mean.

Look up $z = 2$ in Table B.1(a) and read off an *area* of 0.9772. This is the area to the left of $z = 2$, and by symmetry it is also the area to the right of $z = -2$. Since the question requires percentage *less than* 150 cm, we require the area to the left of $z = -2$ i.e.

$1 - 0.9772 = 0.0228$ or 2.28%

Note that the middle of the distribution corresponds to $z = 0$.

To work out (ii), let $x = 200$, and calculate $z = (x - \mu)/\sigma$ again. Now $z = 3$. The area to the left of $z = 3$ is 0.9987, from Table B.1(a). Since the question asks for percentage greater than 200 cm, we require the area to the right of 200, i.e.

$1 - 0.9987 = 0.0013$ or 0.13%

So finally (iii). The percentage between 150 and 200 cm must be:

$100 - 2.28 - 0.13 = 97.59\%$

It is left as an exercise for the reader to use Table B.1(a) to obtain the results quoted immediately before this example, using the same normal distribution example. □

---

* It can the shown that z (calculated in this way) has a normal distribution with a mean of 0 and standard deviation 1, the so called standardised normal distribution (Figure 7.4).

*Example 7.2* 125

The normal distribution, which has been touched on in this section, is useful not only in obtaining probabilities and percentages for a known population, but also because many of the statistical methods used to draw inferences about populations from sample data require the assumption that the variable of interest is at least approximately normal. We need to understand what we are assuming in any statistical method, and more importantly whether that assumption is a valid one for our data.

# 8

# Hypothesis tests and confidence intervals for means

*In recounting Mr Herbert's reforms, Miss Nightingale brought the results of them, after her usual manner, to the statistical test.*

## 8.1 INTRODUCTION

Hypothesis tests and confidence intervals are the names of two aspects of what statisticians call 'statistical inference'. By this they mean 'drawing conclusions from sample data about the populations from which the samples were taken'. In a medical study the patients (or subjects) we select should ideally be a *random* sample from a defined population. If this is not the case it may not be possible to draw any useful conclusions about the population in question.

In the above the word 'population' means all the measurements (or counts) of interest in our study, the word 'sample' means a sub-set of the population, and a 'random sample' is such that each measurement in the population has the same probability of being included in the sample.

### 8.1.1 Example from case study 1

Consider the ages of the 92 men in the control group (see Table 1.1). These 92 ages are a sample. The population is the ages of all males on the list from an Abingdon group practice. We are told that the sample is random.

From the 92 ages we can obtain the following summary statistics:

$\bar{x} = 41.6$ years
$s = 9.6$ years
$n = 92$

The value for $s$ is calculated as follows. We are given $se = 1.0$ in Table 1.1. But

$$se = \frac{s}{\sqrt{n}}$$

(section 7.3), so

$$s = \sqrt{n} \cdot se = \sqrt{92} \times 1.0 = 9.6$$

What do the summary statistics in the above example tell us about the population of ages of men in a group practice in Abingdon? There are two main approaches to answering this question. One is the 'confidence-interval' approach, and the other is the 'hypothesis-test' approach. These will be explained in the following sections with reference to the above and other examples.

### 8.2 CONFIDENCE INTERVAL FOR A POPULATION MEAN (LARGE SAMPLES)

Suppose we ask the question: 'What is the mean age of the population?' The answer is: 'We don't know, since we only have a sample, not the whole population'. Let us call the population mean $\mu$, to distinguish it from the sample mean, $\bar{x}$, which we *do* know if we have taken a random sample. Statistical theory shows that if we calculate the two values of:

$$\bar{x} - \frac{1.96s}{\sqrt{n}} \quad \text{and} \quad \bar{x} + \frac{1.96s}{\sqrt{n}}$$

then we can be '95% confident' that the unknown $\mu$ lies between these '95% confidence limits' as long as $n$ is large ($n > 30$, say). If the sample is not large, a different formula applies under certain conditions (section 8.4).

Notice the form of the above formula, namely:

$$\text{sample estimate} \pm 2se$$

### 8.2.1 Example from case study 1

Using the ages from the example of section 8.1, $n = 92$ is greater than 30. So a 95% 'confidence interval' for $\mu$ is given by:

$$41.6 - \frac{1.96 \times 9.6}{\sqrt{92}} \quad \text{to} \quad 41.6 + \frac{1.96 \times 9.6}{\sqrt{92}}$$

$$41.6 - 2.0 \quad \text{to} \quad 41.6 + 2.0$$

$$39.6 \quad \text{to} \quad 43.6$$

So we are 95% confident that the population mean age lies between 39.6 and 43.6 years.

Similar calculations could be performed for the following three numerical variables in Table 1.1: ages of the other 3 groups (e.g. female control); weights of 4 groups (e.g. female control); heights of 4 groups (e.g. female control).

## 8.3 INTERPRETATION OF A CONFIDENCE INTERVAL

The meaning of the two values which define a 95% confidence interval needs clarifying, as does the meaning of the '95%' and also the word 'confidence'! One interpretation is as follows.

If in our careers as would-be statisticians we calculate lots of 95% confidence intervals, each of which is an attempt to estimate an unknown population parameter such as $\mu$ (the mean), then on 95% of occasions we will 'capture' the true unknown parameter between the two values we state. However, 5% of the time the unknown parameter will lie outside the 95% confidence interval. In any particular case, we will not know whether we have captured the parameter.

The value 95% is a conventional level, a balance between a high percentage value and a reasonably narrow confidence interval. From our discussion in section 7.5 we can obtain confidence intervals with other levels of confidence. A 68.3% confidence interval runs from

$$\bar{x} - \frac{s}{\sqrt{n}} \quad \text{to} \quad \bar{x} + \frac{s}{\sqrt{n}}$$

and a 99.8% confidence interval from

$$\bar{x} - \frac{3.09s}{\sqrt{n}} \quad \text{and} \quad \bar{x} + \frac{3.09s}{\sqrt{n}}$$

for large $n$ in both cases. The first of these would give a narrower interval than for a confidence level of 95%, but we would 'capture' the true unknown value of $\mu$ only 68.3% of the time. We will virtually always capture the mean $\mu$ in the second case, but the 99.8% confidence interval is about 50% wider than the 95% confidence interval.

## 8.4 CONFIDENCE INTERVAL FOR A POPULATION MEAN (SMALL SAMPLES)

Suppose we obtain $\bar{x} = 41.6$ years, $s = 9.6$ years, but for a random sample of only $n = 10$ men in our control group. Clearly $n = 10$ is too small to use the large-sample formula, so that the formula $\bar{x} \pm (1.96s/\sqrt{n})$ does not apply.

These summary statistics may have arisen from the following sample data (figures in years):

27  33  36  37  38  39  45  50  55  56

(It is left as an exercise for the reader to check that $\bar{x} = 41.6$ and $s = 9.6$ years for these data, following the method in section 7.3).

What formula can we use when $n$ is small? The answer is:

$$\bar{x} \pm \frac{ts}{\sqrt{n}}$$

as long as the variable in question is approximately normally distributed. Here $t$ is a number we can obtain simply from Table B.2 at the back of this book, but the sting is in the tail! How do we know our variable is at least approximately normal? We can draw a dot-plot as follows, but we cannot be certain about our conclusion, especially if $n$ is only 10. Ironically, the larger the sample the easier it is to tell whether the distribution is 'normal', but also the larger the sample the less important is the assumption of normality! In the end it is usually a matter of judgement based on experience.

Consider the following five dot-plots:

```
X       X      XXXXXX      X       X

X             XXXXXXXXX             X

X  X  X  X  X  X  X  X  X  X

XXXXX                         XXXXX

XXXXXXX        X        X        X
```

My judgement says that the first three are reasonably approximately normal (they are symmetrical and more or less bunched in the middle), but that the other two are unlikely to be normal. The data in our example can be represented as follows on a dot-plot:

```
   x     x    xxxx   x     x     xx
  25    30    35    40    45    50    55
```

and are approximately normal (in my opinion!).

In order to use the formula $\bar{x} \pm (ts/\sqrt{n})$ we need to be able to look up the correct value of $t$ in Table B.2. This value depends on $\alpha$ and $v$:

- For 95% confidence, $\alpha = (1 - 0.95)/2 = 0.025$.
- $v$ (the Greek lower-case letter 'nu') stands for 'degrees of freedom, and $v = n - 1$ when we use the formula $\bar{x} \pm (ts/\sqrt{n})$ (For a fuller discussion of the topic 'degrees of freedom', see books listed in Appendix A.)

For our example, $t = 2.26$, since $\alpha = 0.025$ for 95% confidence, and $v = n - 1 = 10 - 1 = 9$. So a 95% confidence interval for the population mean age is:

$$41.6 \pm \frac{2.26 \times 9.6}{\sqrt{10}}$$

34.7 to 48.5 years

Notice that this is a much wider interval than obtained for the much larger sample size of 92 (see section 8.2). The moral is the same: larger samples give more information about the population (as one would expect) by providing better estimates (= narrower confidence intervals) of population parameters.

A useful practical tip is to use the formula $\bar{x} \pm (ts/\sqrt{n})$ whether $n$ is small or large (as long as the variable is approximately normal when $n$ is small). This can be justified by choosing a large value of $n$, and hence a large value of $v$, and noting that the value of $t$ at the bottom of the column headed 0.025 in Table B.2 is 1.96.

### 8.5 HYPOTHESIS TEST FOR A POPULATION MEAN (SMALL OR LARGE SAMPLES)

We now set out the hypothesis-test approach to inference for small samples, when we can make the same assumption as in section 8.4, namely that the variable is approximately normal. This method can also be used when we have a large ($n > 30$) sample, but the assumption is no longer needed in that case.

The main idea of a hypothesis test (sometimes called 'a test of significance') is to set up *two* hypotheses about a population parameter, such as the mean $\mu$. We then collect sample data and decide which of the two hypotheses is better supported by the data. For example, if $\mu$ is the mean age of a population, our two hypotheses may be:

$H_0$: $\mu$ = 45 years
$H_1$: $\mu \neq$ 45 years

The first, $H_0$, is called the **null** hypothesis (note the zero subscript) and also indicates the idea of 'no difference' or arithmetically '= 0'. Here $H_0$ implies that there is no difference between $\mu$ and 45. The alternative hypothesis $H_1$, indicates there is 'some difference'.

Here are the seven steps which we should use in any hypothesis test (left-hand side of page) and how these steps apply to the example taken from Table 1.1 for the ages of the men in the control group (right-hand side of page).

| Step No. | Hypothesis test method | Example, testing $H_0$: $\mu$ = 45 |
|---|---|---|
| 1 | Set up a null hypothesis, $H_0$ | $H_0$: $\mu$ = 45 |
| 2 | Set up an alternative hypothesis, $H_1$ | $H_1$: $\mu \neq$ 45 (see note (ii) below) |
| 3 | State the significance level of the test, which is the risk of rejecting $H_0$ when $H_0$ is the correct hypothesis (see note (iii) below). | 5% level of significance |
| 4 | Calculate a test statistic using an appropriate formula. | Calc $t = \dfrac{\bar{x} - 45}{s/\sqrt{n}}$ $= \dfrac{41.6 - 45}{9.6/\sqrt{10}}$ $= -1.12$ (assuming sample data as section 8.4) |
| 5 | Look up tabulated test statistic | Tab $t$ = 2.26 for |

$$\alpha = \frac{\text{Sig. level}}{2} = \frac{0.05}{2}$$

$$= 0.025$$

and $v = n - 1 = 10 - 1 = 9$
(see note ii)

| 6 | Compare calculated and tabulated test statistics | If Calc $t$ is numerically greater than Tab $t$, reject $H_0$ (see note iv) Since $1.12 < 2.26$, $H_0$ is not rejected. |
| 7 | Draw a conclusion | The mean age of men in the control group is not significantly different from 45 (5% level) (see note v) |

*Notes*

(i) The working in the example above applies only if the variable 'age' is approximately normally distributed. We decided this was the case in section 8.4.

(ii) This is called a two-sided $H_1$, since $\mu \neq 45$ implies less than or greater than 45. We will use this type of alternative hypothesis in this book, unless there is a good reason to do otherwise.

(iii) 5% is the conventional level used in hypothesis testing. It is the small risk we have to run in concluding that we should reject $H_0$ when in reality $H_0$ is the correct hypothesis. We should *never* think of 5% as the 'probability that $H_0$ is correct'.

(iv) The 'decision rule' for deciding whether to reject $H_0$ can be thought of graphically as in Figure 8.1. Think of $t$-distributions as a family of shapes similar to the shape of the normal distribution. Each member of the family has the same mean of zero, but different d.f. It is only when the d.f. are large that the $t$ and the normal merge into each other.

Any value of Calc $t$ between $-2.26$ and $+2.26$ leads to non-rejection of $H_0$. But any value 'in the tails', i.e. numerically greater than 2.26, leads to rejection of $H_0$ at the 5% level.

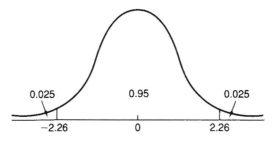

**Figure 8.1** A $t$-distribution with 9 d.f.

(v) Statistical computer packages often give 'p-values' rather than stating whether a null hypothesis is or is not rejected at a particular level of significance. The following statements are equivalent in pairs:

$p > 0.05$ means the same as $H_0$ is not rejected at the 5% level

$p < 0.05$ means the same as $H_0$ is rejected at the 5% level, but by implication, not at the 1% level.

$p < 0.01$ means the same as $H_0$ is rejected at the 1% level, but by implication not at the 0.1% level.

$p < 0.001$ means the same as $H_0$ is rejected at the 0.1% level.

The levels 5%, 1%, 0.1% are conventional levels and are equivalent to p-values of 0.05, 0.01 and 0.001. They are not, of course, the probabilities that '$H_0$ is correct'!

## 8.6 THE CONFIDENCE INTERVAL AND HYPOTHESIS TEST APPROACHES COMPARED

Consider the two approaches in relation to the following sample data and the problem of estimating the mean age of the population.

$\bar{x} = 41.6$ years
$s = 9.6$ years
$n = 10$

We stated in section 8.4 that we were 95% confident that the population mean age is between 34.7 and 48.5 years. We stated in section 8.5 that the mean age of men in the control group is not significantly different from 45 years. These two statements are compatible in the sense that 45 lies in the interval 34.7 to 48.5. It can be shown that any null hypothesis value of the population mean which lies inside the 95% confidence interval will not be rejected at the 5% level of significance.

We can conclude that a confidence interval can be used to test any number of hypotheses, whereas a hypothesis test only indicates whether we should reject a particular hypothesis. In this sense a confidence interval is much more useful than a hypothesis test, which is probably why more and more medical journals insist on the former being quoted.

## 8.7 CONFIDENCE INTERVAL AND HYPOTHESIS TEST FOR COMPARING TWO MEANS (PAIRED t-TEST)

Rather than considering the mean of one population and testing a particular value in a hypothesis test, it is much more common in practice to compare the means of two populations in order to decide whether the two means are significantly different. This is equivalent to testing the null hypothesis $H_0$: $\mu_1 = \mu_2$ where $\mu_1$ and $\mu_2$ are the two population means, or to deciding whether the confidence interval for $\mu_1 - \mu_2$ contains the value zero. If it does, we would not reject $H_0$.

Before we discuss this (in section 8.8) we need to consider problems which apparently involve two populations, but which do in fact reduce to one-population problems. In this case the methods of sections 8.4 and 8.5 apply. Such problems occur when we deal with 'paired' data, which naturally arise for example, in the following cases:

- The same patients (or subjects) are measured both before and after some treatment.
- The same subjects are measured in two independent sites, each receiving different treatments.
- An intervention group receives a treatment and a control group, matched in pairs with the intervention group, receives no treatment.

All these give rise to data which can be tabulated in the following form, for which we are much more interested in the *differences* between pairs of treatment responses than in the individual treatment responses.

| Subject No | 1 | 2 | 3 | 4 | 5 |
|---|---|---|---|---|---|
| Before | x | x | x | x | x |
| After | x | x | x | x | x |
| *Difference* | *x* | *x* | *x* | *x* | *x* |
| Subject No | 1 | 2 | 3 | 4 | 5 |
| Site 1 | x | x | x | x | x |
| Site 2 | x | x | x | x | x |
| *Difference* | *x* | *x* | *x* | *x* | *x* |
| Pair No | 1 | 2 | 3 | 4 | 5 |
| Intervention | x | x | x | x | x |
| Control | x | x | x | x | x |
| *Difference* | *x* | *x* | *x* | *x* | *x* |

### 8.7.1 Example from case study 6

Consider the comparison referred to in row 1 of Table 6.4, and refer to Figure 6.1. It is clear that 37 children received the treatment:

adenoidectomy + myringotomy                      for ear 1
adenoidectomy + myringotomy + grommet     for ear 2.

The differences in the audiometry results between the ears for the child were clearly of interest here, so we have paired data. Seven weeks after the operation the mean difference (presumably ear 2 − ear 1) is stated as 8.1, with a confidence interval of 3.0 to 13.3.

If we use $\bar{d}$ (instead of $\bar{x}$) to stand for the sample mean difference in dB), $s_d$ (instead of $s$) to stand for the standard deviation of $d$), and $\mu_d$ (instead of $\mu$) to stand for the population mean difference), then we have $\bar{d} = 8.1$. The formula $\bar{x} \pm (ts/\sqrt{n})$ from section 8.4 becomes

$$\bar{d} \pm \frac{ts_d}{\sqrt{n}}$$

giving 3.0 to 13.3 as a 95% confidence interval for $\mu_d$.

We know $n < 37$ here because we are told 'audiometric data were not obtained on every occasion in about 15% of children (22/147 is about 15%). Let us suppose $n = 37$ less 15% = 31. Hence $t = 2.04$ (Table B.2), and solving

$$\frac{ts_d}{\sqrt{n}} = 8.1 - 3.0 = 5.1$$

gives

$$s_d = \frac{5.1\sqrt{31}}{2.04} = 13.9$$

So a summary of the 31 pairs of data could have been:

$$\bar{d} = 8.1$$
$$s_d = 13.9$$
$$n = 31$$

This 'detective work' is not, however, necessary to agree with the conclusion reached from the result 8.1 (3.0 to 13.3) quoted in Table 6.4, because this 95% confidence interval does *not* contain the value zero, and since both 3.0 and 13.3 are positive, we can conclude that 'the mean effect of the grommet (once adenoidectomy and myringotomy have been carried out) is significant at the 5% level' or as stated in the case study, '$p < 0.05$'.

Of course, we should have considered any assumptions here, namely whether the differences between the audiometry readings of the ears of the subjects were approximately normal. But we have a sample of 37, reduced to 31 possibly, so we have a large enough sample to ignore this assumption.

The following is a check on our conclusions, and is included to show the formal steps of a paired *t*-test:

1 $H_0$: $\mu_d = 0$. The differences come from a population with a mean of zero, which means that the grommet has no effect (once adenoidectomy and myringotomy have been carried out).

2 $H_1$: $\mu_d \neq 0$. The grommet has some effect.

3 5% level.

4 Calc $t = \dfrac{\bar{d} - 0}{s_d/\sqrt{n}} = \dfrac{8.1}{13.9/\sqrt{31}} = 3.24$.

5 Tab $t = 2.04$ for $\alpha = \dfrac{0.05}{2} = 0.025$, $\nu = 31 - 1 = 30$ (Table B.2).

6 Since $3.24 > 2.042$, reject $H_0$.

7 The mean effect of the grommet is significant (5% level). Clearly, since 8.1 is positive the effect of the grommet is to *increase* audiometry levels. This agrees with the conclusion drawn already by reference only to the 95% confidence interval for the mean difference.

## 8.8 CONFIDENCE INTERVAL AND HYPOTHESIS TEST FOR COMPARING TWO MEANS (UNPAIRED *t*-TEST)

In medical studies, in which two populations are compared, paired data are encountered much less often than unpaired data. We usually wish to compare two populations, for example, intervention versus control, by taking independent samples from each population. There is often no question of pairing an observation from one sample with an observation from the other sample. (This might be the case if the intervention group were matched one-for-one with the control group.)

### 8.8.1 Examples from the case studies

In case study 1, baseline characteristics are compared between the control and intervention groups, separately for men and women (Table 1.1). Table 1.5 shows results of 40 unpaired *t*-tests.

In case study 3, Table 3.1 shows results and the commentary indicates five unpaired *t*-tests. The commentary also indicates the results of two unpaired *t*-tests comparing consultation rates for men and women.

In case study 6, Table 6.5 gives the results of 42 unpaired *t*-tests.

We will choose one example only from these to illustrate this very popular method.

Using the information in Table 1.1, we obtain a 95% confidence interval for $\mu_1 - \mu_2$, the difference in the mean ages of the men in the control and intervention groups, and test whether the difference in mean ages is significantly different from zero (at the 5% level). The sample data are as follows:

| *Control* | *Intervention* |
|-----------|----------------|
| $\bar{x}_1 = 41.6$ | $\bar{x}_2 = 42.1$ |
| $s_1 = 9.6$ | $s_2 = 9.8$ |
| $n_1 = 92$ | $n_2 = 97$ |

where $s_1$ and $s_2$ are calculated as in section 8.1.1.

Instead of $\bar{x} \pm (ts/\sqrt{n})$, the formula for the confidence interval for a population mean, we need another formula for the confidence interval for $\mu_1 - \mu_2$, the difference between two population means. The formula is:

$$(\bar{x}_1 - \bar{x}_2) \pm ts \sqrt{\frac{1}{n_1} + \frac{1}{n_2}}$$

where

$$s^2 = \frac{(n_1 - 1)s_1^2 + (n_2 - 1)s_2^2}{n_1 + n_2 - 2}$$

We need to be able to make two assumptions in using this formula:

- The variable, in this case age, is approximately normally distributed in both populations. However, since $n_1$ and $n_2$ are both large, we can ignore this assumption.
- The population variances, $\sigma_1^2$ and $\sigma_2^2$ are equal. We will assume this is true in this case since the sample variances $s_1^2$ and $s_2^2$ are in good agreement here (however, this point will be discussed in greater depth in section 9.5).

In the formula above, $s^2$ is our estimate of the 'common' variance of the two populations, based on information from both samples. So, for our example,

$$s^2 = \frac{91 \times 9.6^2 + 96 \times 9.8^2}{92 + 97 - 2} = 94.15$$

giving $s = 9.70$. Also, in the above, $t$ is from Table B.2 for $\alpha = 0.025$, if we want a 95% confidence interval, and $v = n_1 + n_2 - 2 = 187$. So $t = 1.96$ approximately since 187 is very large. Hence a 95% confidence interval for $(\mu_1 - \mu_2)$ is

$$(41.6 - 42.1) \pm 1.96 \times 9.70\sqrt{\frac{1}{92} + \frac{1}{97}}$$
$$-0.5 \pm 2.8$$
$$-3.3 \text{ to } 2.3$$

Since this interval contains the value zero, we would *not* wish to reject $H_0: \mu_1 = \mu_2$ in favour of $H_1: \mu_1 \neq \mu_2$ at the 5% level of significance. We could conclude that: the mean ages of the men in the control and intervention groups are not significantly different (5% level).

We could have reached the same conclusion starting with the same sample data, but proceeding by the seven-step hypothesis-test method, as follows:

1 $H_0: \mu_1 = \mu_2$
2 $H_1: \mu_1 \neq \mu_2$
3 5% level

4 Calc $t = \dfrac{\bar{x}_1 - \bar{x}_2}{s\sqrt{\dfrac{1}{n_1} + \dfrac{1}{n_2}}}$, where $s^2 = \dfrac{(n_1 - 1)s_1^2 - (n_2 - 1)s_2^2}{n_1 + n_2 - 2}$.

So $s = 9.70$ as before, and

$$\text{Calc } t = \frac{41.6 - 42.1}{9.70\sqrt{\dfrac{1}{92} + \dfrac{1}{97}}} = \frac{-0.5}{1.41} = -0.35$$

5 Tab $t = 1.96$, for $\alpha = \dfrac{0.05}{2} = 0.025$, $v = n_1 + n_2 - 2 = 187$

6 Since $0.35 < 1.96$, do not reject $H_0$.

7 Mean ages of men in the control and intervention groups are not significantly different. This agrees with case study 1 since there is no asterisk for this comparison in Table 1.1, implying that $p > 0.05$.

The assumptions needed for this test are identical to those needed for the calculations of a 95% confidence interval for $\mu_1 - \mu_2$.

## 8.9 PRACTICAL AND STATISTICAL SIGNIFICANCE

In designing case study 6 the authors decided that they thought that 10 dB should be the minimum difference in levels of hearing between treatments that might be regarded as 'clinically important' and that the trial should have a 95% chance of detecting such a difference between the two treatments at the 5% level of significance. (This kind of information can be used to help decide the study size, as we shall discuss in Chapter 14.) The phrase 'clinically important' means the same as 'practically significant', as opposed to the phrase 'statistically significant' which, if stated in the context of a 5% level, means that there is a 5% chance that the null hypothesis has been wrongly rejected, that is to say, there is a 5% chance of concluding that there is some difference between the means, say, while in reality there is no difference.

However, the fact of concluding that there is a statistically significant difference between the means does not make this difference practically important. It is possible to choose such large sample sizes that any difference in sample means can be shown to be statistically significant. On other occasions the difference between two sample means may appear to be practically important but cannot be shown to be statistically significant because, for example, the sample size may be too small.

## 8.10 WHAT IF THE ASSUMPTIONS OF THE METHODS IN THIS CHAPTER ARE NOT VALID?

The simple answer is that the methods should not be applied. There are alternative methods, particularly tests called nonparametric tests, for which the assumptions are much less severe than for the corresponding $t$-tests (Rees 1989).

# 9

# More on comparing means, the analysis of variance and the F-test

*The common lilac flowers, according to Quetelet's law, when the sum of squares of the mean daily temperatures, counted from the end of the frosts, equals 4264° centigrade.*

## 9.1 INTRODUCTION

The methods of section 8.8 are suitable when we wish to compare two means, the data are unpaired, and assumptions of normality and equality of variance are justified. Sometimes we wish to compare more than two means, in which case a technique called *analysis of variance* (ANOVA) is useful. In its simplest form the assumptions required in ANOVA are the same as those for the unpaired *t*-test. ANOVA is in fact a general method which can be applied in the analysis of data from a variety of designed experiments. This chapter will provide an introduction to ANOVA.

### 9.1.1 Example from case study 3

In Table 3.6, mean consultation rates are compared for four groups of women at six points in time (groups of men are similarly compared). The four groups are labelled N/N, A/A, A/N and N/A, where N/A, for example, means achieving a total GHQ score of ≤8 at the beginning and >8 at the end of the follow-up year (see Table 3.4).

We are told in Table 3.6 that the mean consultation rates for women in the study year were as follows:

| N/N | A/A | A/N | N/A |
|-----|-----|-----|-----|
| 6.2(59) | 13.2(10) | 9.3(20) | 8.0(7) |

where the numbers in brackets are the numbers of patients. We could use the mathematical notation:

| N/N | A/A | A/N | N/A |
|-----|-----|-----|-----|
| $\bar{x}_1(n_1)$ | $\bar{x}_2(n_2)$ | $\bar{x}_3(n_3)$ | $\bar{x}_4(n_4)$ |

We are told that $p < 0.01$ was obtained when these data were analysed by ANOVA. Notice that we are not given any standard deviations ($s_1$, $s_2$, $s_3$, $s_4$), so there is no way we can check the conclusion $p < 0.01$ from this case study. Two questions are worth asking at this stage. The first question is what null hypothesis was tested as part of the ANOVA. The answer is $H_0: \mu_1 = \mu_2 = \mu_3 = \mu_4$, that is, the mean consultation rates are the same for the four populations. The second question is the title of the next section.

## 9.2 WHY NOT CARRY OUT $T$-TESTS TO COMPARE THE MEANS IN PAIRS?

Why did the authors of case study 3 not test $H_0: \mu_1 = \mu_2$, using an unpaired $t$-test, and repeat for the other five pairs of means? The answer is twofold:

- The six $t$-tests would not be independent, since, for example, the data from each sample would be used in three tests.
- The risk of wrongly rejecting at least one of the six null hypotheses would be much larger than 5%.

So, we need another approach to analysing these data, and this is where ANOVA is useful.

## 9.3 A NUMERICAL EXAMPLE OF ONE-WAY ANOVA

We cannot analyse the data provided in Table 3.6, because we are not given enough information. However, here is a similar example with fictitious raw data, in which the number of subjects is **balanced**, that is, the same in each group. (It is not difficult to analyse unbalanced ANOVAs like the one in case study 3.)

| Groups | 1 | | 2 | | 3 | | 4 | | |
|---|---|---|---|---|---|---|---|---|---|
| | 6 | | 11 | | 8 | | 6 | | |
| | 7 | | 14 | | 9 | | 7 | | |
| | 7 | | 17 | | 11 | | 9 | | |
| | 8 | | 16 | | 12 | | 6 | | |
| | 5 | | 14 | | 4 | | 5 | | |
| | 4 | | 10 | | 7 | | 8 | | |
| | 3 | | 9 | | 8 | | 12 | | |
| | 2 | | 12 | | 6 | | 11 | | |
| | 6 | | 13 | | 9 | | 9 | | |
| | 9 | | 15 | | 10 | | 10 | | |
| Sum | 57 | + | 131 | + | 84 | + | 83 | = | 355 |
| Sum of Squares (unadjusted) | 369 | + | 1777 | + | 756 | + | 737 | = | 3639 |
| Mean | 5.7 | | 13.1 | | 8.4 | | 8.3 | | |
| Standard deviation | 2.2 | | 2.6 | | 2.4 | | 2.3 | | |
| No. of patients | 10 | | 10 | | 10 | | 10 | | |

The numbers in the main part of the table are the number of consultations per year for 40 patients, 10 in each group. The summary statistics shown are either useful in carrying out ANOVA or ones we would normally calculate.

The basis of the ANOVA here is as follows:

We consider all 40 observations and calculate what is called the **total sum of squares**:

$$3639 - \frac{355^2}{40} = 488.4$$

We then use only the group summary statistics and calculate what is called the **between-groups sum of squares**:

$$\frac{57^2}{10} + \frac{131^2}{10} + \frac{84^2}{10} + \frac{83^2}{10} - \frac{355^2}{40} = 284.9$$

The difference between these is called the **within-groups sum of squares**:

$$488.4 - 284.9 = 203.5$$

Each of the above sums of squares is associated with its own degrees of freedom. Since 40 observations are considered, the **total d.f.** is

$$40 - 1 = 39$$

Since four groups are considered, the **between-groups d.f.** is

$$4 - 1 = 3$$

The difference between these two answers is called the **within-groups d.f.**:

$$39 - 3 = 36.$$

From these six calculations the ANOVA table can now be formed, and this leads to the hypothesis test of

$H_0: \mu_1 = \mu_2 = \mu_3 = \mu_4$

| Source of variation | SS | d.f. | MS | Calc. F-ratio |
|---|---|---|---|---|
| Between groups | 284.9 | 3 | 94.97 | 16.8 |
| Within groups | 203.5 | 36 | 5.65 | |
| Total | 488.4 | 39 | | |

Columns 1–3 have already been explained. MS stands for **mean square**, and the entries are obtained by dividing each SS value by the corresponding d.f. value. Finally, the calculated F-ratio is the ratio of the two mean squares.

Here are the seven steps of the hypothesis test, followed by a discussion of the assumptions needed in order for the test to be valid:

1 $H_0: \mu_1 = \mu_2 = \mu_3 = \mu_4$.
2 $H_1$: at least two of the means differ.
3 5% level.
4 Calc $F$ = 16.8 from above.
5 Tab $F$ = 2.87 for (3, 36) d.f. using 5% F-tables (Table B.3). Three degrees of freedom are associated with the top of the F-ratio calculation. Look along the top of the F-tables for the number 3. Look down the side for the other d.f., here 36. Hence Tab $F$ = 2.87 (interpolating between 30 and 40 d.f.).
6 Since 16.8 > 2.87, reject $H_0$.
7 At least two of the means are significantly different.

Two assumptions have been made here:

● Each sample comes from an approximately normally distributed population. To check this objectively is the same problem as previously encountered in section 8.4. A more subjective check is to draw four dot-plots, one per group, and this reveals at least approximate normality.

- Each sample comes from a population with the same variance ($\sigma_1^2 = \sigma_2^2 = \sigma_3^2 = \sigma_4^2$). Since the sample variances are $2.2^2$, $2.6^2$, $2.4^2$ and $2.3^2$, which are in good agreement, we can assume this assumption is justified.

If we think that either of these assumptions is not valid, we should not carry out the $F$-test. A non-parametric test which may be used, and which requires less stringent assumptions, is the Kruskal–Wallis test (see Altman 1991).

### 9.4 A POSTERIOR TEST AFTER ANOVA

If we have decided not to reject the null hypothesis in ANOVA, that is the end of the analysis because 'the means are not significantly different'. However, if we reject the null hypothesis all we know is that there are some differences between some means, but which means are significantly different? One way of refining the conclusions of ANOVA when some significance has been established is to use a posterior test, one of which is called the SNK test (after its authors Student, Neuman and Keuls). Here is how it works on the example of the previous section.

The means are ranked in increasing order of magnitude,

| Group | 1 | 4 | 3 | 2 |
|-------|-----|-----|-----|------|
| Mean  | 5.7 | 8.3 | 8.4 | 13.1 |
| Rank  | 1   | 2   | 3   | 4    |

The two means which differ most are compared by dividing their difference by *se*, where:

$$se = \sqrt{\frac{\text{Within-group MS}}{\text{No. of observations per group}}} = \sqrt{\frac{5.65}{10}} = 0.75$$

So we calculate

$$\frac{13.1 - 5.7}{0.75} = 9.8$$

If this is greater than the value of $q$ from Table B.4, then the corresponding means are significantly different at the 5% level.

Then the two means which differ most among the remaining pairs are compared, and so on. Here are the results of all possible comparisons:

| Comparison | Difference in means | Diff. in mean $\div$ se | Tab conclusion q | |
|---|---|---|---|---|
| Rank 1 (G1) v Rank 4 (G2) | 13.1 − 5.7 = 7.4 | 9.8 | 3.81 | Sig. diff. |
| Rank 2 (G4) v Rank 4 (G2) | 13.1 − 8.3 = 4.8 | 6.4 | 3.46 | Sig. diff. |
| Rank 3 (G3) v Rank 4 (G2) | 13.1 − 8.4 = 4.7 | 6.3 | 2.87 | Sig. diff. |
| Rank 1 (G1) v Rank 3 (G3) | 8.4 − 5.7 = 2.7 | 3.6 | 3.46 | Sig. diff. |
| Rank 1 (G1) v Rank 2 (G4) | 8.3 − 5.7 = 2.6 | 3.5 | 2.87 | Sig. diff. |
| Rank 2 (G4) v Rank 3 (G3) | 8.4 − 8.3 = 0.1 | 0.1 | 2.87 | Not sig. |

Only the difference between the means of groups 3 and 4 is not significant. So we can think of three distinct sets of means: the mean for group 1 is the lowest; the means for groups 3 and 4 are the next lowest; and the mean for group 2 is the highest.

This concludes the analysis of these data. Notice that in Table 3.6, we are told that $p < 0.01$ (for the women in the study year) but we are not told of any posterior tests that have been carried out. Hence the authors were entitled to say only that at least two of the four means are significantly different (5% level).

## 9.5 THE F-TEST TO COMPARE TWO VARIANCES

Another use of the $F$-test is to test the hypothesis $H_0$: $\sigma_1^2 = \sigma_2^2$, where $\sigma_1^2$ and $\sigma_2^2$ are the variances of two populations. This is exactly the test we needed in section 8.8, so we will use the example described there (from Table 1.1) to illustrate this test.

The sample data were

$s_1 = 9.6$ $s_2 = 9.8$
$n_1 = 92$ $n_2 = 97$

We obtain

$$\text{Calc } F = \frac{s_1^2}{s_2^2} \quad \text{or} \quad \frac{s_2^2}{s_1^2}$$

whichever is the larger, and compare it with Tab $F$, using Table B.3. Here are the 7 steps needed:

1 $H_0$: $\sigma_1^2 = \sigma_2^2$.
2 $H_1$: $\sigma_1^2 \neq \sigma_2^2$.
3 5% level.
4 Calc $F = \dfrac{s_2^2}{s_1^2} = \dfrac{9.8^2}{9.6^2} = 1.04$.

5 Tab $F > 1.25$ for (96, 91) d.f.*
6 Since $1.04 < 1.25$; do not reject $H_0$.
7 The variances are not significantly different (5% level).

This means that one of the assumptions of the $t$-test used in section 8.8 is justified. However, the $F$-test itself requires the assumption that the two populations are normally distributed.

---

* Table B.3 cannot tell us the answer here, because the table is not extensive enough. However, it can be shown that Table $F = 1.25$ for ($\infty$, 120) d.f. (see Altman 1991). It follows that Tab $F > 1.25$ for (96, 91) d.f.

# Hypothesis tests and confidence intervals for percentages

*In the China expedition every required arrangement for the preservation of health was made, with the result that the mortality of this force, including wounded, was little more than 3 per cent per annum. During the first seven months of the Crimean War the mortality was at a rate of 60 per cent per annum from disease alone, a rate which exceeds that of the Great Plague in London.*

## 10.1 INTRODUCTION

We saw in Chapter 7 how some variables of interest in a medical study are numerical and some are non-numerical. With the former we will probably wish to calculate summary statistics such as the mean and standard deviation for groups of patients and compare the means using the inferential methods of Chapters 8 and 9.

For non-numerical variables our summary statistics are likely to be in the form of percentages (see section 7.4 for examples). However, the inferential questions of interest are very similar. For example:

- If we know the percentage who fall into a certain category of a non-numerical variable for a random sample of patients, what does this tell us about the population percentage?
- If we compare two populations by taking a random sample from each, and observe the percentages falling into a certain category of a non-numerical variable, what does

this tell us about the difference between the population percentages?

As before we have the confidence-interval approach and the hypothesis-test approach, and the realization that the two approaches are related.

## 10.2 CONFIDENCE INTERVAL FOR A PERCENTAGE

### 10.2.1 Examples from case study 2

We are told that, of 642 controls, 5.3% did not smoke at the one-month follow-up; the 95% confidence interval for the population percentage is quoted as from 3.6% to 7.0%. How were 3.6% and 7.0% determined?

The formula used is as follows:

$$\frac{x}{n} \pm 1.96 \sqrt{\frac{\frac{x}{n}\left(1 - \frac{x}{n}\right)}{n}}$$

Here $n$ is the number of patients (or subjects), so $n = 642$, and $x$ is the number of patients (or subjects) who fall into the category of interest, so $x = 34$ (5.3% of 642).

Notice how a 95% confidence interval contains the number 1.96 (recall section 8.2) and the idea of sample estimate $\pm 2se$, so here we have the standard error of proportion

$$\sqrt{\frac{\frac{x}{n}\left(1 - \frac{x}{n}\right)}{n}}$$

The formula only holds if $x > 5$ and $(n - x) > 5$, both clearly true here. If we now feed in $x = 34$, $n = 642$, we obtain:

$$0.053 \pm 1.96 \sqrt{\frac{0.053 \times 0.947}{642}}$$

0.0357 to 0.070
or 3.6% to 7.0% (as quoted above).

So we are 95% confident that in the population of controls the percentage not smoking at the one-month follow-up lies between 3.6 and 7.0%. The interpretation of this is similar to that given in section 8.3.

There are in all 24 confidence intervals quoted in Tables 2.1 and 2.2, and lots more in Table 6.6. The conditions $x > 5$ and $(n - x) > 5$ are clearly true for all the examples in Tables 2.1 and 2.2.

## 10.3 HYPOTHESIS TEST FOR A PERCENTAGE

We could invoke the idea of section 8.6 and conclude that any null hypothesis which specifies a value for the population percentage outside the 95% confidence interval will be rejected at the 5% level of significance, while any value inside the interval will not be rejected. However, this argument does not hold exactly in certain marginal cases, so it will be useful to go through the formal steps of the hypothesis test.

### 10.3.1 Example from case study 1

We use the data of the previous section to test the null hypothesis that, in the population, $p = 0.1 \, (= 10\%)$, where $p$ stands for the proportion of the population who do not smoke at the one-month follow-up (see Table 1.1).

1 $H_0$: $p = 0.1$.
2 $H_1$: $p \neq 0.1$.
3 5% level.
4 Calc $z = (x/n - 0.1)/\sqrt{[0.1(1 - 0.1)]/n}$ provided in this that $0.1n > 5$ and $n(1 - 0.1) > 5$. Since $n = 642$, both conditions are easily satisfied. Since $x = 34$,

$$\text{Calc } z = \frac{0.053 - 0.1}{\sqrt{\dfrac{0.1 \times 0.9}{642}}} = -3.97$$

5 Tab $z = 1.96$ (see Figure 10.1 or see Table B.1(b)).
6 Since $3.97 > 1.96$, reject $H_0$, see Figure 10.1 and Note (i).
7 The proportion of controls who did not smoke at the one-month follow-up was significantly different from 0.1 (=10%), at the 5% level of significance.

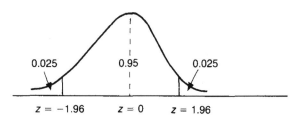

**Figure 10.1** The standard normal distribution

*Notes*

(i)  The decision rule used here is similar to the one described
     in section 8.5. Here we use a normal rather than a
     *t*-distribution, and reject $H_0$ if Calc $z$ falls in the tails of the
     distribution, i.e. $>1.96$ or $<-1.96$, but do not reject $H_0$ if
     Calc $z$, lies between $-1.96$ and $+1.96$.

(ii) We could quote $p < 0.05$, since $3.97 > 1.96$. We could also
     quote $p < 0.01$, since $3.97 > 2.58$ ($z = \pm 2.58$ implies 0.005
     in each tail, see Table B.1(b)). We could, finally, quote $p <$
     0.001, since $3.97 > 3.29$ ($z = \pm 3.29$ implies 0.0005 in each
     tail, see Table B.1(b)).

### 10.4 CONFIDENCE INTERVAL FOR THE DIFFERENCE BETWEEN TWO PERCENTAGES

Just as there is usually more interest in comparing two means
than comparing one mean with some hypothesized value, the
same is true with percentages. We usually want to compare the
percentage of one group of patients who fall into a certain
category with the percentage of another group who fall into the
same category.

Although the hypothesis-test approach using the $\chi^2$ (chi-
square) test is extremely popular in medical journals, we will
first of all discuss the confidence-interval approach because we
have seen in section 8.6 that a confidence interval contains
much more information than a hypothesis test.

### 10.4.1 Example from case study 2

I will not choose examples from Table 2.1 to compare the
control and intervention groups, because of the complication

there in which attenders and non-attenders have to be lumped together using a weighted average method before the comparison can be made.

Instead, I will compare the attenders and non-attenders in the intervention group, using the data for the one-month follow up. Of 751 attenders, 10.9% (i.e. 82) did not smoke at the one-month follow-up. Of 367 non-attenders, 6.5% (i.e. 24) did not smoke at the one-month follow-up. Calculate a 95% confidence interval for the population difference in the percentage who did not smoke at one month follow up.

The formula we need is:

$$\frac{x_1}{n_1} - \frac{x_2}{n_2} \pm 1.96 \sqrt{\frac{\frac{x_1}{n_1}\left(1 - \frac{x_1}{n_1}\right)}{n_1} + \frac{\frac{x_2}{n_2}\left(1 - \frac{x_2}{n_2}\right)}{n_2}}$$

The formula only holds if $x_1 > 5$, $(n_1 - x_1) > 5$, and if $x_2 > 5$, $(n_2 - x_2) > 5$. Here $n_1 = 751$, $x_1 = 82$, $n_2 = 367$ and $x_2 = 24$, so our conditions are all clearly true.

Thus a 95% confidence interval is:

$$0.109 - 0.065 \pm 1.96 \sqrt{\frac{0.109 \times 0.901}{751} + \frac{0.065 \times 0.935}{367}}$$

$0.044 \pm 0.034$
$0.010$ to $0.078$
or $1.0$ to $7.8\%$

We are 95% confident that the difference in the two population percentages lies between 1.0% and 7.8%. Using the idea of section 8.6, we would expect to reject the null hypothesis that this difference is 0%, that is, we would expect to reject the null hypotheses that there is no difference between the population percentages, because 0% lies outside the interval 1.0% to 7.8%. We will see if we reach this conclusion in the next section using the hypothesis-test approach.

## 10.5 HYPOTHESIS TEST FOR THE DIFFERENCE BETWEEN TWO PERCENTAGES

While it is possible to use an extension of the $z$-test described in section 10.3 to test the null hypothesis that two population percentages are equal, a method much more favoured by medical journals is the $\chi^2$ (chi-square) test. The two approaches can be shown to lead to exactly the same conclusion, so I will describe only the latter method here.

### 10.5.1 Examples from case studies

*Case study 2*

Using the data from section 10.4 to compare attenders and non-attenders at one-month follow-up, we can draw up a two-way 'contingency table' (Table 10.1). The important features of this table are:

- The entries are frequencies (not percentages or proportions).
- The entries are independent of each other, so that each observation (there are 1118) is from a different individual. Nor are matched pairs of individuals involved.

The method involves calculating the frequencies we would expect if the null hypothesis (that the population percentage of non-smokers is the same for attenders and non-attenders) were true. We then compare the expected frequencies ($E$) with those we have observed ($O$) (see Table 10.1) and calculate the test statistic:

$$\text{Calc } \chi^2 = \sum \frac{(|O - E| - \frac{1}{2})^2}{E}$$

where the $\Sigma$ implies summing over all the cells in the contingency table. So the null hypothesis, $H_0$, is that the percentages of non-smokers is the same for attenders and non-attenders. How can we calculate the $E$ (expected) frequencies assuming $H_0$ is true?

Ignoring attendance for the moment, the proportion of non-

**Table 10.1** Numbers smoking and non-smoking at one-month follow-up

|  | Attenders | Non–attenders | Total |
|---|---|---|---|
| Non-smoking | 82 | 24 | 106 |
| Smoking | 669 | 343 | 1012 |
| Total | 751 | 367 | 1118 = Grand Total |

smokers is 106/1118. If this applies equally to attenders and non-attenders, then the expected number of attenders who are non-smokers is

$$\frac{106}{1118} \times 751 = 71.2$$

This expected frequency corresponds to an observed frequency ($O$) of 82 for non-smoking attenders. Looking at the way we calculated the $E$-value above, we can see we (in effect) applied the formula

$$E = \frac{\text{row total} \times \text{column total}}{\text{grand total}}$$

We can apply the same formula to the other three cells in the table to give the following, where the $E$-values are shown in brackets.

|              | Attenders    | Non-attenders |      |
| ------------ | ------------ | ------------- | ---- |
| Non-smoking  | 82 (71.2)    | 24 (34.8)     | 106  |
| Smoking      | 669 (679.8)  | 343 (332.2)   | 1012 |
|              | 751          | 367           | 1018 |

*Notes*

(i) An expected frequency is a sort of average, not a frequency which could occur on a particular occasion. For this reason it does not have to be a whole number. One decimal place is recommended.

(ii) The expected frequencies sum to the same row and column totals as the observed frequencies.

So we now calculate as follows:

$$\text{Calc } \chi^2 = \sum \frac{(|O - E| - \frac{1}{2})^2}{E}$$
$$= \frac{(|82 - 71.2| - \frac{1}{2})^2}{71.2} + \frac{(|24 - 34.8| - \frac{1}{2})^2}{34.8} +$$
$$\frac{(|669 - 679.8| - \frac{1}{2})^2}{679.8} + \frac{(|343 - 322.2| - \frac{1}{2})^2}{332.2}$$

$$= \frac{10.3^2}{71.2} + \frac{10.3^2}{34.8} + \frac{10.3^2}{8} + \frac{10.3^2}{332.2}$$

$$= 5.01$$

where $|O - E|$ means the magnitude of the difference between $O$ and $E$, ignoring the sign. So $|82 - 71.2| = 10.8$, $|24 - 34.8| = 10.8$ and so on.

Here is the complete seven-step hypothesis test:

1 $H_0$: The percentage of non-smokers is the same for the attenders and non-attenders.
2 $H_1$: The percentage of non-smokers is not the same for attenders and non-attenders.
3 5% level.
4 Calc $\chi^2 = 5.01$, from above.
5 Tab $\chi^2 = 3.84$, from Table B.5 for $\alpha = $ sig. level $= 0.05$, and $v = (r - 1)(c - 1) = (2 - 1)(2 - 1) = 1$, where $v$ is the number of degrees of freedom.
6 Since Calc $\chi^2 > $ Tab $\chi^2$, reject $H_0$.
7 The percentages of non-smokers for attenders and non-attenders are significantly different (5% level). Since more attenders are non-smokers than expected ($82 > 71.2$), we can conclude that the percentage of non-smokers for attenders is significantly *greater* than for attenders.

*Notes*

(i)   The formula used here (which incorporates what is known as Yates's correction) applies only to $\chi^2$ tests where there is 1 d.f. The use of Yates's correction is not universal but is recommended by the experts. In all other cases the following simpler formula should be used:

$$\text{Calc } \chi^2 = \sum \frac{(O - E)^2}{E}$$

(ii)  Neither the formula used in this test nor the simpler version should be used if $E$-values are too small. A conservative rule is 'all $E$-values should be at least 5'. This point will be discussed further in examples below.

(iii) For a contingency table with $r$ rows and $c$ columns, the degrees of freedom are:

d.f. $= (r - 1)(c - 1)$

For a 2 × 2 table, d.f. = 1, so Yates's correction should always be used for this size of table.

(iv) Another way of expressing the null hypothesis is to say that attendance and smoking are independent. This is why the $\chi^2$ test is sometimes called the '$\chi^2$ test of independence'.

(v) Instead of 'significant at the 5% level', we could have stated $p < 0.05$. However, $p < 0.01$ is not justified since Tab $\chi^2 = 6.63$ for $\alpha = 0.01$ and $v = 1$, and $5.01 < 6.63$.

(vi) The conclusion reached in this test is the one we expected to reach using the confidence interval approach; see the last paragraph of section 10.4.

## Case study 1

From Table 1.1, 2 × 2 tables can be formed for eight cases. For example, to test whether, for men, the percentages in social class 1 or 2 are the same for the control and intervention groups.

Another twelve 2 × 2 tables can be formed from Table 1.2, all leading to the conclusion $p < 0.001$, so Calc $\chi^2$ must have been greater than 10.83 (Table B.5 for $\alpha = 0.001$, $v = 1$) in each case.

## Case study 3

Table 3.2 is a 3 × 2 table, hence $(3 - 1)(2 - 1) = 2$ d.f. One $E$-value is marginally below 5, but this has been ignored in this case study (see example below on data in Table 3.3).

Table 3.4 is another 4 × 2 table, with all $E$-values above 5.

Each row of Table 3.5 can be used to draw up a 2 × 4 table. The rows of the first table (corresponding to row 1 of Table 3.5) would be labelled 'Given psychiatric diagnosis' and 'Not given psychiatric diagnosis', and the observed frequencies would be:

| | | | |
|-----|----|----|----|
| 13 | 7 | 19 | 6 |
| 107 | 10 | 16 | 8 |
| 120 | 17 | 35 | 14 |

Table 3.3 is a 4 × 2 table, but no fewer than three expected values are below 5.

Let drug 1 be anti-depressants and drug 2 benzodiazepine anxiolytics. Then the table, including expected values in brackets, becomes:

|          | Men       | Women      | Total            |
|----------|-----------|------------|------------------|
| Drug 1   | 9 (7.6)   | 12 (13.4)  | 21               |
| Drug 2   | 4 (7.6)   | 17 (13.4)  | 21               |
| Both     | 4 (3.6)   | 6 ( 6.4)   | 10               |
| Neither  | 3 (1.1)   | 0 ( 1.9)   | 3                |
| Total    | 20        | 35         | 55 = Grand Total |

Applying the conservative rule that all $E$-values should be at least 5, a $\chi^2$-test is *not* appropriate here. But what can be done about small $E$-values?

For a table which is larger than a 2 × 2 table, it *may* be sensible to combine rows and/or columns to make a smaller table in which the expected values are not too small. A 2 × 2 table cannot be collapsed, but another test, the Fisher exact test can be used in this case (section 11.1).

In the case of Table 3.3, clearly we cannot combine the columns since there are only two. Also row 4 (= neither) cannot sensibly be combined with any other row. The only sensible idea is to collapse rows 1, 2 and 3 and obtain:

|                | Men        | Women     | Total            |
|----------------|------------|-----------|------------------|
| Drug taken     | 17 (18.9)  | 35 (33.1) | 52               |
| No Drug taken  | 3 (1.1)    | 0 (1.9)   | 3                |
| Total          | 20         | 35        | 55 = Grand Total |

However, two $E$-values are still below 5, so the $\chi^2$-test is still inappropriate. As mentioned above, the Fisher exact test should be used instead.

There is one more type of table which sometimes occurs. This is when there are more than two rows (or columns) and the categories defining the rows (or columns) are in a natural rank order. This occurred in Table 2.1, where the order

    controls      non-attenders     attenders

is a kind of rank order in the sense that 'on control days, nothing further was done beyond usual care' and 'on intervention days, smokers were asked to make an appointment for a health check' – but some did not attend. The standard $\chi^2$ test does not use this ranking information, but the $\chi^2$ trend test does. This will also be discussed in the next chapter by reference to this particular example.

# 11

---

# Further tests for percentages

*Without supposing, then, that she had solved the ultimate riddle of the universe, Miss Nightingale had hold of an hypothesis which solved for her many of her mediate riddles.*

## 11.1 THE FISHER EXACT TEST

As mentioned in the last chapter, this test is appropriate when the frequencies can be set out in a $2 \times 2$ contingency table, but at least one of the expected frequencies is less than 5.

Suppose the observed frequencies are $a$, $b$, $c$, $d$ as follows:

|       |       |       | Total |
|-------|-------|-------|-------|
|       | $a$   | $b$   | $a + b$ |
|       | $c$   | $d$   | $c + d$ |
| Total | $a + c$ | $b + d$ | $n =$ Grand Total |

Then the method is to calculate the following probability:

$$\text{probability} = \frac{(a + b)! \, (c + d)! \, (a + c)! \, (b + d)!}{n! \, a! \, b! \, c! \, d!}$$

The procedure is repeated for all similar tables, having the same marginal totals, and for those tables having a probability less than or equal to the probability for the initial table. Then the total probability is calculated.

If the usual null hypothesis is adopted, namely independence between the row and column variables, then the null hypothesis is rejected at the 5% level of significance if the calculated value of the probability is *less than 0.05*. (This decision rule is contrary to most if not all the others seen in this book, where a high calculated value leads to the rejection of the null hypothesis. This test is different because we calculate the probability of obtaining the data we have collected on the assumption that the null hypothesis is correct. So a *low* calculated probability throws doubt on the null hypothesis.)

### 11.1.1 Example from case study 3

Table 3.3 can be collapsed to form a 2 × 2 table (already discussed in section 10.5).

|                | Men | Women | Total |
|----------------|-----|-------|-------|
| Drugs taken    | 17  | 35    | 52    |
| No drugs taken | 3   | 0     | 3     |
| Total          | 20  | 35    | 55    |

Here $a = 17$, $b = 35$, $c = 3$, $d = 0$, $n = 55$; so

$$\text{probability} = \frac{52! \times 3! \times 20! \times 35!}{55! \times 17! \times 35! \times 3! \times 0!} = 0.0435$$

There are three other tables which can be drawn up with the same marginal totals, but all have probabilities greater than 0.0435. Hence the calculated probability is 0.0435.

The seven-step procedure is as follows:

1 $H_0$: The percentages of men and women taking drugs is the same.
2 $H_1$: The percentages of men and women taking drugs are different.
3 5% level.
4 Calc probability = 0.0435.
5 There is no 'tabulated' probability, but 0.05 is the 'critical' level of probability.
6 Since $0.0435 < 0.05$, the null hypotheses is rejected.
7 There is a significantly higher percentage of women taking drugs (5% level).

Note that, as for the $\chi^2$ tests of Chapter 10, the 55 observations must be independent.

## 11.2 $\chi^2$ (CHI-SQUARE) TREND TEST

This test was introduced at the end of section 10.5, as follows.

### 11.2.1 Example from case study 2

For the final column of Table 2.1, Since

0.9% of 642 is  6 (must be a whole number)
3.3% of 367 is 12 (must be a whole number)
4.7% of 751 is 35 (must be a whole number)

the following 2 × 3 table can be formed:

|  | Controls | Non-attenders | Attenders | Total |
|---|---|---|---|---|
| No. with sustained cessations of smoking | 6 | 12 | 35 | 53 |
| No. without sustained cessations of smoking | 636 | 355 | 716 | 1707 |
| Total | 642 | 367 | 751 | 1760 |
| Score, $x$ | −1 | 0 | 1 |  |

(the three scores −1, 0, 1 imply a linear trend). We now calculate

$$6 \times -1 + 12 \times 0 + 35 \times 1 = 29$$
$$642 \times -1 + 367 \times 0 + 751 \times 1 = 109$$
$$642 \times (-1)^2 + 367 \times 0^2 + 751 \times 1^2 = 1393$$

and then

$$\text{Calc } \chi_1^2 = \frac{1760 \{1760 \times 29 - 53 \times 109\}^2}{53 \times 1707 \{1760 \times 1393 - 109^2\}}$$
$$= \frac{3.6058 \times 10^{12}}{2.2073 \times 10^{11}}$$
$$= 16.34$$

Compared with Tab $\chi^2$ for 1 d.f. of 3.84 ($p = 0.05$), 6.63 ($p = 0.01$) and 10.83 ($p = 0.001$), we can conclue '$p < 0.001$' (since $16.3 > 10.83$) as quoted in the Results section of case study 2.

However, we can do a little more analysis to expand on these conclusions as follows. If we calculate the standard Calc $\chi^2$ from the 2 × 3 table above, we obtain expected frequencies:

| 19.3 | 11.1 | 22.6 |
|------|------|------|
| 622.7 | 355.9 | 728.4 |

and hence

$$\text{Calc } \chi_2^2 = \sum \frac{(O - E)^2}{E}$$

$$= \frac{(6 - 19.3)^2}{19.3} + \frac{(12 - 11.1)^2}{11.1} + \frac{(35 - 22.6)^2}{22.6} +$$

$$\frac{(636 - 622.7)^2}{622.7} + \frac{(355 - 355.9)^2}{355.9} + \frac{(716 - 728.4)^2}{728.4}$$

$$= 16.54 \text{ with } (2 - 1)(3 - 1) = 2 \,\text{d.f.}$$

The test for departure from a linear trend gives

$$\text{Calc } \chi_3^2 = \chi_2^2 - \chi_1^2 = 16.54 - 16.34 = 0.20$$

This is associated with 1 d.f. and is clearly not significant at the 5% level $(0.20 < 3.84)$.

The conclusion of a linear trend is reinforced – it is in fact the only significant effect to be observed here – and the conclusion seen in the Results section of case study 2 that 'the rate of sustained cessation in the non-attenders was intermediate to the rate in controls and attenders' is confirmed.

# 12

# Regression and correlation

## 12.1 INTRODUCTION

So far we have not considered problems in which we observe two (or more) numerical variables on a number of subjects (or patients), but this kind of problem is quite common. For example, we might be interested in both the heights and weights of a sample of subjects. We could draw a scatter diagram with, say, height on the vertical axis and weight on the horizontal axis. If there appears to be a linear relation between the two variables, we could do two analyses:

- Obtain the equation of the best-fit* line for our data, of the form:

$$\text{height} = a + b \times \text{weight} \qquad (12.1)$$

- Calculate the correlation coefficient between the variables height and weight.

Equation (12.1) is called the simple linear regression equation of 'height on weight', and would be useful if our aim was to *predict* the height of a subject from the weight of the subject. The numerical values of $a$ and $b$ can be calculated directly from the heights and weights of the sample of subjects.

In the second type of analysis, the aim is to *measure* the degree to which the two variables are linearly related. The value of the correlation coefficient can also be tested to see whether we should reject the null hypothesis that there is no correlation in the bivariate population of heights and weights.

As in some of the statistical methods discussed in earlier chapters, there are some assumptions we need to check on before we can go ahead with the required analysis.

---

* Best here means that the sum of squares of distances from the points to the line in the vertical direction is minimized.

Generalizing equation (12.1), we can postulate a *simple* linear regression equation of the form:

$$y = a + bx$$

where $x$ is sometimes called the *independent* or *explanatory* variable, and $y$ is called the *dependent* variable.

If we think that a number of explanatory variables $x_1, x_2, \ldots$ might be useful in predicting the dependent variable, we can use a *multiple* linear regression equation, which takes the form:

$$y = a + b_1x_1 + b_2x_2 + \ldots$$

In this chapter we will cover simple linear regression and correlation, and introduce the main ideas involved in multiple regression analysis, all by reference to examples based on case study 4.

## 12.2 SIMPLE LINEAR REGRESSION

### 12.2.1 Example for the case study 4

In Figure 4.1(a) we see a scatter diagram relating the dependent variable $y$, the number of prescriptions per patient, to the explanatory variable $x$, the SMR, for 98 family practitioner committees in 1987 (these are the 98 'subjects'). Since we are not given the individual values of $x$ and $y$ for each 'subject' and because 98 is a large number, I have selected at random ten 'subjects' and read off the approximate values of $x$ and $y$ from Figure 4.1(a) to give the following raw data (graphed in Figure 12.1).

For these ten data points, the slope ($b$) and intercept ($a$) of the linear regression line are given by the following:

$$b = \frac{\Sigma xy - n\bar{x}\bar{y}}{\Sigma x^2 - n\bar{x}^2} = \frac{7403.0 - 10 \times 97.5 \times 7.53}{95\,493 - 10 \times 97.5^2} = \frac{61.25}{430.5} = 0.142$$

$$a = \bar{y} - b\bar{x} = 7.53 - 0.142 \times 97.5 = -6.34$$

So the simple linear regression equation is:

$$y = -6.34 + 0.142x$$

|  | Number of prescriptions per patient (y) | Standardized mortality ratio (x) |
|---|---|---|
|  | 11.2 | 112 |
|  | 6.7 | 87 |
|  | 6.5 | 99 |
|  | 6.7 | 96 |
|  | 8.0 | 92 |
|  | 8.3 | 102 |
|  | 7.5 | 91 |
|  | 6.7 | 98 |
|  | 7.1 | 101 |
|  | 6.6 | 97 |
| Sums | $\Sigma y = 75.3$ | $\Sigma x = 975$ |
| Sums of squares | $\Sigma y^2 = 585.47$ | $\Sigma x^2 = 95\,493$ |
| Sums of products | $\Sigma xy = 7403.0$ | |
| Number of subjects | $n = 10$ | |

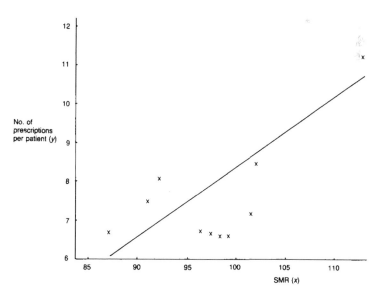

**Figure 12.1** A scatter diagram of prescription rate against standardized mortality ratio

To draw the linear regression line, proceed as follows. First, choose $x = 87$ (the smallest value of $x$). This gives a predicted value of $y$ of:

$$-6.34 + 0.142 \times 87 = 6.04.$$

Plot (87, 6.04). Then choose $x = 112$ (the largest value of $x$). This give a predicted value of $y$ of:

$$-6.34 + 0.142 \times 112 = 9.60.$$

Plot (112, 9.60). Join the two plotted points. This is the regression line. Roughly half the points should be above the line and half below the line.

### 12.3 CORRELATION COEFFICIENT

Pearson's product moment correlation coefficient, $r$ (Pearson's $r$ for short) can be calculated for samples of bivariate data, and measures the degree to which two variables are linearly related. The formula for $r$ is:

$$r = \frac{\Sigma xy - n\bar{x}\bar{y}}{\sqrt{[\Sigma x^2 - n\bar{x}^2][\Sigma y^2 - n\bar{y}^2]}}$$

### 12.3.1 Example from case study 4

Using the random sample of data used in the previous section, we find that:

$$r = \frac{7403.0 - 10 \times 97.5 \times 7.53}{\sqrt{[95\,493 - 10 \times 97.5^2][585.47 - 10 \times 7.53^2]}}$$

$$= \frac{61.25}{\sqrt{430.5 \times 18.461}}$$

$$= 0.687$$

It can be shown that $r$ must lie in the range $-1$ to $+1$.

```
                                    0.687
_____×_____
  -1                  0                      +1
```

If there is a trend such that, as $x$ increases $y$ increases, $r$ will be positive. If there is a trend such that, as $x$ increases $y$

decreases, $r$ will be negative. If there is no trend one way or the other, the value of $r$ will be close to zero. If all the points lie on a straight line $r$ will be $+1$ or $-1$, depending on the trend.

Clearly the points in our scatter diagram do exhibit a positive trend, but they do not all lie on a straight line. The value $r = 0.687$ can be described as 'reasonably high positive correlation'.

If we know, as we do in this case, that the sample taken is a random one from a population (in this case the population of all 98 committees) we can test the hypothesis that the population correlation coefficient $\rho$ is zero.

Here are the 7 steps required:

1 $H_0$: $\rho = 0$ No correlation between $x$ and $y$ in the population.
2 $H_1$: $\rho \neq 0$ Some correlation between $x$ and $y$ in the population.
3 5% level.

4 Calc $t = r\sqrt{\dfrac{n-2}{1-r^2}}$

$\quad = 0.687\sqrt{\dfrac{10-2}{1-0.687^2}}$

$\quad = 2.67$

5 Tab $t = 2.31$ for $\alpha = \dfrac{\text{Sig. level}}{2} = \dfrac{0.05}{2} = 0.025$

and $v = (n-2) = 10 - 2 = 8$ (Table B.2).

6 Since $2.67 > 2.31$, we reject $H_0$.
7 There is a significant correlation between $x$ and $y$ (5% level). Clearly since $r$ is positive, we can also conclude that there is significant *positive* correlation between $x$ and $y$.

The assumptions needed for this test are that $x$ and $y$ are both normally distributed. Dot-plots of $x$ and $y$ show these assumptions are not unreasonable.

### 12.4 ANOVA APPLIED TO SIMPLE LINEAR REGRESSION ANALYSIS

The technique of analysis of variance (ANOVA) was introduced in Chapter 9 by reference to Table 3.6. In that example, we decided how much of the total variation in the number of consultations per year could be attributed to variation between groups and how much to variation within groups.

We can apply the same kind of idea to simple linear regression analysis by asking how much of the total variation in the dependent variable $y$, can be attributed to the explanatory variable $x$, and how much of the variation is unexplained.

In order to answer this question we need to calculate the total, regression and residual sums of squares, respectively, and their corresponding degrees of freedom.

### 12.4.1 Example from case study 4

Let us once again use the random sample of ten family practitioner committees, where $y$ is the number of prescriptions per patient, and $x$ is SMR.

$$
\begin{aligned}
\text{Total sum of squares} \quad &= \Sigma y^2 - n\bar{y}^2 \\
&= 585.47 - 10 \times 7.53^2 \\
&= 18.46
\end{aligned}
$$

$$
\begin{aligned}
\text{Regression sum of squares} &= b^2(\Sigma x^2 - n\bar{x}^2) \\
&= 0.142^2(95\,493 - 10 \times 97.5^2) \\
&= 8.72
\end{aligned}
$$

$$
\begin{aligned}
\text{Residual sum of squares} \quad &= 18.46 - 8.72 \\
&= 9.74
\end{aligned}
$$

Each of the above sums of squares is associated with certain degrees of freedom. Since there are $n = 10$ data points, the total d.f. is $n - 1 = 9$. Since there is only one $x$ variable, the regression d.f. is 1. The difference, the residual d.f., is therefore 8.

The ANOVA table can now be formed, and this leads to the hypothesis test of $H_0$: $\beta = 0$, where $\beta$ is the slope of the 'population' regression line, that is, the line we would draw through the points on the scatter diagram for the whole population.

| Source of variation | SS | d.f. | MS | Calc F ratio |
|---|---|---|---|---|
| Regression (on $x$) | 8.72 | 1 | 8.72 | 7.15 |
| Residual | 9.74 | 8 | 1.22 | |
| Total | 18.46 | 9 | | |

Columns 1–3 have already been explained. MS stands for mean square, and the entries are obtained by dividing each SS

value by the corresponding d.f. value. Finally, the calculated *F*-ratio is the ratio of the two mean squares.

Here are the seven steps of the hypothesis test, followed by a discussion of the assumptions needed in order for the test to be valid.

1 $H_0$: $\beta = 0$ The population regression line is horizontal.
2 $H_1$: $\beta \neq 0$ The population regression line is not horizontal.
3 5% level.
4 Calc $F = 7.15$, from above.
5 Tab $F = 5.32$ for (1, 8) d.f. using 5% $F$ tables. One degree of freedom is associated with the *top* of the $F$ ratio calculation. Look down the side for the other d.f., here 8. Hence Tab $F = 5.32$.
6 Since $7.15 > 5.32$, reject $H_0$.
7 The slope is significantly *different* from zero (5% level). Clearly, since $b = 0.142$, the slope is significantly *greater* than zero (5% level).

Whenever we make any inference (either a confidence interval or a hypothesis test) in regression analysis, the same assumptions apply. The residuals are approximately normally distributed about the line (in the $y$ direction) with zero mean and constant variance. This is really three assumptions in one, and needs elaborating. We define the residuals to be the distances from the points to the line in the $y$ direction. For our example, the ten residuals are easily calculated, using the formula:

residual = actual $y$ − predicted $y$

| x | Actual y | Predicted y | Residual |
|-----|----------|-------------|----------|
| 112 | 11.2 | 9.6 | 1.6 |
| 87 | 6.7 | 6.0 | 0.7 |
| 99 | 6.5 | 7.7 | −1.2 |
| 96 | 6.7 | 7.3 | −0.6 |
| 92 | 8.0 | 6.7 | 1.3 |
| 102 | 8.3 | 8.2 | 0.1 |
| 91 | 7.5 | 6.6 | 0.9 |
| 98 | 6.7 | 7.6 | −0.9 |
| 101 | 7.1 | 8.0 | −0.9 |
| 97 | 6.6 | 7.5 | −0.9 |

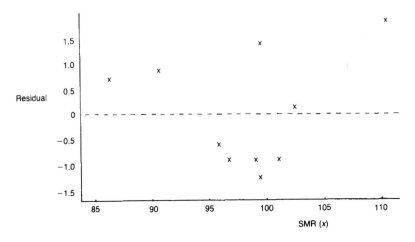

**Figure 12.2** Plot of residuals against SMR

It is useful to plot the residuals against $x$ as a graphical check on the three assumptions. This is done in Figure 12.2. We see, firstly, that the residuals appear to have a mean close to zero. Secondly, the variation about the line is approximately the same for all values of $x$. Thirdly, it is impossible to say whether the distribution of residuals is approximately normal, when there are only ten data points, although 'bunching in the middle' is not exactly in evidence here.

### 12.5 HOW USEFUL IS THE REGRESSION EQUATION?

#### 12.5.1 Example from case study 4

How useful is the equation relating $y$, the number of prescriptions per patient, to $x$, the SMR, obtained from the random sample of ten 'subjects'? Another way of asking this question is to ask: 'How much of the variation in $y$ is 'explained' by the variable $x$?'

The answer, in terms of sums of squares from the ANOVA table is, $8.72/18.46 = 0.47$. This ratio is called the **coefficient of determination**, $R^2$. Thus 47% of the variation in $y$ is 'explained by' $x$.

When there is only one explanatory variable $x$ in a regression equation, as is the case here, $R^2$ is numerically equal to $r^2$, the square of the correlation coefficient between $x$ and $y$ (looking back to section 12.3 we see that $r^2 = 0.687^2 = 0.47 = R^2$, as expected).

However the idea of the coefficient of determination can be extended to multiple linear regression analysis where there are two or more explanatory variables, whereas this is not true for the correlation coefficient.

## 12.6 INTRODUCTION TO MULTIPLE REGRESSION ANALYSIS

In case study 4 a number of explanatory variables in addition to SMR are studied to see if a multiple linear regression equation is better at predicting prescription rates than one involving SMR alone. We hope to aid understanding of how this was done by adding in one further explanatory variable, namely Jarman score, and seeing its effect. To the data in section 12.2 we now add a third column $x_2$, having relabelled $x$ as $x_1$:

| No. of prescriptions per patient (y) | Standardized mortality ratio ($x_1$) | Jarman score ($x_2$) |
|---|---|---|
| 11.2 | 112 | 35 |
| 6.7 | 87 | −15 |
| 6.5 | 99 | 10 |
| 6.7 | 96 | 0 |
| 8.0 | 92 | −10 |
| 8.3 | 102 | 20 |
| 7.5 | 91 | −15 |
| 6.7 | 98 | 10 |
| 7.1 | 101 | 5 |
| 6.6 | 97 | 5 |

It should be emphasized here that it was impossible to obtain the Jarman scores even approximately from case study 4 for the random sample of ten family practitioner committees. The scores in the table above are entirely fictitious but they have been 'fiddled' to be very highly positively correlated with SMR in order to illustrate a common problem in multiple linear regression called 'multicollinearity', to be dicussed below!

The next step might be to plot a scatter diagram of $y$ against $x_2$ (Figure 12.3) to see:

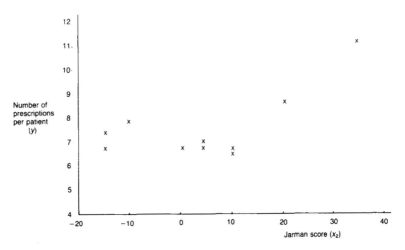

**Figure 12.3** A scatter diagram of prescription rate against Jarman score

- whether it exhibits a linear trend;
- whether $x_2$ is better than $x_1$ at explaining the variation in $y$.

We see that the trend is arguably linear, showing a positive correlation, but the trend is clearly not perfectly linear. To decide whether $x_2$ is better than $x_1$ at explaining the variation in $y$, we can either calculate the correlation coefficient between $y$ and $x_2$ or draw up an ANOVA table. We will do both, but the calculations are left as an exercise for the reader. For $y$ and $x_2$, $r$ = 0.625 (less than 0.687 for $y$ and $x_1$). The ANOVA table in as follows:

| Source of variation | SS | d.f. | MS | Calc F-ratio |
|---|---|---|---|---|
| Regression (on $x_2$) | 7.22 | 1 | 7.22 | 5.14 |
| Residual | 11.24 | 8 | 1.41 | |
| Total | 18.46 | 9 | | |

For $y$ and $x_2$, $R^2$ = 7.22/18.46 = 0.39 (= $0.625^2$, of course). So $x_1$ is better than $x_2$ at explaining the variation in since $R^2$ = 0.47 for $x_1$, but only 0.39 for $x_2$.

The next question we will ask is: 'If we put $x_1$ *and* $x_2$ into the regression equation, will $x_1$ and $x_2$ together explain $(0.47 + 0.39)100 = 86\%$ of the variation in $y$?' The answer would be 'yes' if we could have chosen $x_1$ and $x_2$ to be completely uncorrelated, or if they were by chance completely uncorrelated; otherwise no.

We can easily find out how much of the variation in $y$ is explained by $x_1$ and $x_2$ together by drawing up another ANOVA table, this time done by computer (since the calculations are too complicated to be done on a calculator.

| Source of variation | SS | d.f. | MS | Calc F-ratio |
|---|---|---|---|---|
| Regression on $x_1$ and $x_2$ | 9.06 | 2 | 4.53 | 3.38 |
| Residual | 9.40 | 7 | 1.34 | |
| Total | 18.46 | 9 | | |

The new value of $R^2$ for $x_1$ and $x_2$ is $9.06/18.46 = 0.49$, which is larger than 0.47 when $x_1$ alone was considered, but only just, and much smaller than the 0.86 we would like to have seen. The reason for this very modest increase is that $x_1$ and $x_2$ are themselves highly correlated (for $x_1$ and $x_2$, the correlation coefficient is 0.964).

Another point here is that $R^2$ is bound to increase as more and more explanatory variables are added into the regression equation. For this reason the coefficient of determination can be calculated using a different formula to provide what is called an 'adjusted $R^2$'. Instead of

$$R^2 = \frac{\text{Regression SS}}{\text{Total SS}} = 1 - \frac{\text{Residual SS}}{\text{Total SS}}$$

we use

$$\text{Adjusted } R^2 = 1 - \frac{\text{Residual SS}/(n - p)}{\text{Total SS}/(n - 1)}$$

where $p$ is number of explanatory variables $+1$. For our example, when $x_1$ is in the regression equation, $R^2 = 0.47$, Adj $R^2 = 0.41$; when $x_1$ and $x_2$ are both in the regression equation, $R^2 = 0.49$, Adj $R^2 = 0.35$. It seems that a regression equation in $x_1$ alone is preferred.

Another way of looking at the same problem is to form a

modified version of the last ANOVA table, by splitting up line
1 into regression due to $x_1$ and regression due to $x_2|x_1$,
where $|x_1$ means 'having fitted $x_1$':

| Source of variation | SS | d.f. | MS | Calc F-ratio |
|---|---|---|---|---|
| Regression on $x_1$ | 8.72 | 1 | 8.72 | 6.51 |
| Regression on $x_2\|x_1$ | 0.34 | 1 | 0.34 | 0.25 |
| Residual | 9.40 | 7 | 1.34 | |
| Total | 18.46 | 9 | | |

Both $F$-ratios are for 1 and 7 d.f., for which Tab $F = 5.59$, from
Table B.3. We can conclude that $x_1$ explains a significant
amount of the variation in $y$ (since $6.51 > 5.59$). We also
conclude that, once $x_1$ is in the equation, $x_2$ does not explain a
significant amount of that variation in $y$ which is so far unexplained by $x_1$ (since $0.25 < 5.59$). Again we conclude that an
equation in $x_1$ alone is preferred.

Finally, in this section, we will look at the effect of multicollinearity on the coefficients of the explanatory variables in the
three regression equations of (i) $y$ on $x_1$, (ii) $y$ on $x_2$, (iii) $y$ on $x_1$
and $x_2$. These equations (only the first of which we have seen
before in section 12.2) are as follows:

(i) $y = 6.34 + 0.142x_1$ ($x$ was relabelled $x_1$ in section 12.6)
(ii) $y = 7.27 + 0.057x_2$
(iii) $y = -16.1 + 0.245x_1 - 0.0468x_2$

In (i) we interpret 0.142 to be 'the increase in $y$ when $x_1$ is
increased by 1 unit'. But notice in (iii) how the coefficient of $x_1$
has increased dramatically. The reason for that is the high
correlation between $x_1$ and $x_2$. We cannot simply introduce one
variable ($x_1$ or $x_2$) without altering the coefficient of the other
($x_2$ or $x_1$) in this case. Similarly, the coefficient of $x_2$ varies
greatly from (ii) to (iii).

## 12.7 HOW GOOD IS OUR PREDICTED VALUE OF $Y$ IN SIMPLE LINEAR REGRESSION ANALYSIS?

We introduced this chapter by saying that regression analysis
may be appropriate if we wish to *predict* a dependent variable
from an explantory variable.

### 12.7.1 Example from case study 4

We found the linear regression equation of $y$ on $x$ to be

$$y = -6.34 + 0.142x$$

Suppose we wish to predict the value of $y$ when x = 100. The simple answer is

$$-6.34 + 0.142 \times 100 = 7.86.$$

But how good is this prediction? Another way of answering this question is 'What is the 95% confidence interval for the mean value of $y$ (the number of prescriptions per patient) for family practitioner committees having a value of $x$ (SMR) of 100?' (Refer to Chapter 8 if necessary to revise confidence intervals.)

The formula to use to obtain a 95% confidence interval for a predicted value of $y$ when $x = x_0$, say, where the simple linear regression equation is $y = a + bx$, is

$$(a + bx_0) \pm ts\sqrt{\frac{1}{n} + \frac{(x_0 - \bar{x})^2}{\Sigma x^2 - n\bar{x}^2}}$$

where $t$ is obtained from Table B.2 for $\alpha = 0.025$, $v = n - 2$, and $s$, the residual standard deviation is obtained from ANOVA, since $s = \sqrt{\text{Residual MS}}$.

For our example, $a = -6.34$, $b = 0.142$, $x_0 = 100$, $t = 2.31$ for $\alpha = 0.025$ and $v = 10 - 2 = 8$, and $s = \sqrt{1.22} = 1.10$. So a 95% confidence interval for $y$ when $x = 100$ is:

$$(-6.34 + 0.142 \times 100) \pm 2.31 \times 1.10 \sqrt{\frac{1}{10} + \frac{(100 - 97.5)^2}{95\,493 - 10 \times 97.5^2}}$$

$$7.86 \pm 0.86$$
$$7 \text{ to } 8.7$$

So we are 95% confident that the mean number of prescriptions per patient for family practitioner committees having an SMR of 100 will be between 7.0 and 8.7. We can calculate confidence intervals for other values of $x$ other than 100. If we do we will notice that the '± term', i.e. the amount we have to add on and subtract gets larger as we move to either relatively large or small values of $x$. The confidence interval is narrowest when $x = 97.5$, the average value of $x$ for our data.

## 12.8 CORRELATION, FURTHER DISCUSSION

Earlier in this chapter correlation coefficients were used to indicate the strength of the linear association between pairs of variables. This is their main use, but in journals and newspapers they are sometimes used for other, more dubious, reasons. We read of 'a significant correlation between variables $x$ and $y'$ as though this indicated that if we could move $x$ in the required direction, then this would have the effect of changing $y$ in its required direction.

Simplistic economic arguments abound concerning the correlation between, say, the number of people unemployed and bank interest rates, as though these variables were perfectly correlated and no other variables exerted any effect on either.

We have all heard of nonsense correlations, for example if one takes the average GP's salary and the number of thefts of cars for each year in the last decade, the ten point scatter diagram might show a positive correlation, which could be (incorrectly) tested for statistical significance (incorrectly because the last 10 years are hardly a random sample from a population). Even if it were correct to do this test, nobody would seriously suggest that reducing GP's salaries would reduce car theft.

We can also be misled by (a) non-linear correlations, (b) correlations when either variable is artificially restricted. So the scatter diagram showing the time to run 100 metres for a random sample of individuals (Figure 12.4) might be non-linear over the age range 5–65 years (Figure 12.4(a)), but could be linear over the range 35–65 years (Figure 12.4(b)).

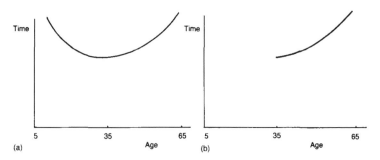

**Figure 12.4** Time to run 100 metres against age of runner

The best advice is to plot your data first on a scatter diagram, calculate the correlation coefficient if you must, test it for statistical significance (it must be from a random sample, and both variables should be normally distributed), think very carefully about making statements about 'cause' and 'effect'. Other advice is to suspect any correlation coefficient in a medical journal, unless it is used simply as part of a regression analysis. Even then inference should only be made using data from a random sample, and not (as in case study 4) using data from the whole population (of family practitioner committees).

# 13

# Sensitivity and specificity

## 13.1 DEFINITIONS

Suppose we wish to use a diagnostic test to help us decide whether a patient has a particular condition. We will assume that the result of the test is either positive or negative, where 'positive' is a diagnosis that the patient *does* have the condition, while 'negative' is a diagnosis that the patient *does not* have the condition. Suppose further that, *in reality*, the condition is either *present* or *absent* in the patient. We can now conceive of four types of situation, corresponding to the four categories in the following table:

|  | Condition | |
|---|---|---|
|  | Absent | Present |
| Diagnostic − | True Negatives | False Negatives |
| Test | (*a*) | (*b*) |
| Indication + | False Positives | True Positives |
|  | (*c*) | (*d*) |

Notice that positive (+) and negative (−) refer to the diagnostic test indication, while true and false refer to whether the test indication agrees or does not agree with the real state of the patient. This notation is used by most texts, but by no means all, so be warned!

Let *a, b, c, d* refer to the numbers in each category, assuming the test has been carried out on *n* patients, where $n = a + b + c + d$. Then the sensitivity and specificity of the diagnostic test are defined as follows:

$$\text{sensitivity} = \frac{100d}{b + d}$$

This is the percentage of those who have the condition and are correctly diagnosed.

$$\text{specificity} = \frac{100a}{a + c}$$

This is the percentage of those who do not have the condition and are correctly diagnosed.

Common sense tells us that a diagnostic test should have both high sensitivity and specificity. However, in practice it is often found that if we increase one of these measures, the other decreases. A balance must be struck.

Other measures can be computed from the four-category table, for example $100(b + c)/n$, the misclassification rate. The advantage of sensitivity and specificity over a number of other measures is that they are independent of the prevalence of the condition, where

$$\text{prevalence} = \frac{100(b + d)}{n}$$

## 13.2 EXAMPLE FROM CASE STUDY 5

Case study 5 is the only case study which describes the topics of this chapter in any detail. There the general health questionnaire (GHQ) provided the diagnostic test. A score of 9 or more ($\geq 9$) corresponded to a positive test result, while a score of 8 or fewer ($< 9$) corresponded to a negative test result.

The 'condition' of the patient was 'conspicuous and hidden psychiatric morbidity' as assessed by a general practitioner. Presence and absence of the condition are referred to as psychiatric case/not a psychiatric case in Table 5.2. So the GHQ score is being used as a proxy for a doctors assessment of psychiatric morbidity. The following shows a redesigned Table 5.2 for male data only:

|                   | Doctor's Assessment | |
|-------------------|:-------------------:|:-------------------:|
|                   | Not a case (absent) | A case (present) |
| GHQ <9(−)         | 79 (a)              | 7 (b)            |
| Score ≥9(+)       | 11 (c)              | 18 (d)           |

So for men, with a 'cut-off score' of 9,

$$\text{Sensitivity} = \frac{100 \times 18}{7 + 18} = 72\%$$

$$\text{Specificity} = \frac{100 \times 79}{79 + 11} = 88\%$$

The corresponding figures for females are 66% and 87%, respectively.

Figure 5.1 indicates that the specificity of the test could be increased further by increasing the cut-off score, but only at the expense of a lower sensitivity.

There is no objective way of deciding where the balance between the two measures should be struck. For this particular case, the authors of CSV conclude that 'a cut-off point of 9 is best suited for clinical use in general practice'. The reason they give is that 'in general practice it would be more clinically useful for the test to have high specificity (that is relatively few false positives), thus excluding patients whose symtoms are so mild that no therapeutic action is called for'. It is not the purpose of this book to argue for or against this conclusion, but simply to provide an example of how the balance between sensitivity and specificity might be made in a particular instance.

# 14

# Study design

*The struggle was long and arduous; the victory of 1867 was only partial, and indeed there are other parts of her [FN's] designs which even to this day [1913] await fruition.*

## 14.1 INTRODUCTION

Study designs may be thought of as falling into two broad categories: **observational surveys** and **designed experiments**.

In an observational survey, data are collected from a number of patients or subjects by one of the following methods:

- by looking at historical records (**retrospective survey**)
- by looking at current records (**cross-sectional survey**)
- by following up patients/subjects over a period of time in the future (**prospective survey**).

In an observational survey, no intervention treatment is given, where treatment here means 'application of medical care or attention'. The patients or subjects are simply observed as they are.

In contrast, a designed experiment is one in which a treatment is given to a group of patients, the response to the treatment being measured over a period of time in the future. Hence designed experiments are prospective. A particular type of designed experiment used commonly in medicine is the 'randomized controlled trial' (RCT) in which the whole group of patients is divided randomly into a number of sub-groups, each sub-group receiving a different treatment. One of the treatments may be a placebo given to a control group. In an RCT the design is usually prospective to include follow-up of the patient after treatment.

**Table 14.1** Classification of case-study designs

| Case study | Observational survey | | Prospective | Designed experiment (R.C.T.) |
| --- | --- | --- | --- | --- |
| | *Retrospective* | *Cross-sectional* | | |
| 1 | | | | x |
| 2 | | | | x |
| 3 | | | x | |
| 4 | x | | | |
| 5 | | x | | |
| 6 | | | | x |

Using the above ideas our six case studies can be classified as shown in Table 14.1. In case study 1 the 'treatment' was dietary instruction, given to the intervention group, while no advice was given to a control group of subjects.

In case study 2 the 'treatment' was anti-smoking advice, but some who were invited to make an appointment did not attend and received no advice. Again there was a control group. In addition, the attenders who received advice were divided into those who did or did not receive carbon monoxide (CO) monitoring.

In case study 3, a follow-up study, the main aim was to compare the change over a period in the general health questionaire score of patients with the rate at which they consulted their GP, and the psychotropic drugs they were prescribed over the same period.

In case study 4, a retrospective study, historical data were used to try to establish a statistical model relating two 'responses', namely prescription rates and prescription costs in 98 family practitioner committees to a number of 'explanatory' variables such as standardized mortality ratio, Jarman score, age–sex structure and the number of GPs per 1000 population.

In case study 5, a cross-sectional study, the aim was to establish a cut-off point in the general health questionnaire score which would give acceptable sensitivity and specificity in deciding whether or not a patient would, according to a GP, be assessed 'as a psychiatric case'.

Finally, in case study 6 in there were five treatment surgeries for glue ear, all aimed at improving hearing in patients. Patients were followed up for two years after surgery.

This section is only the briefest introduction to study design, and the next section is also a short one on study size. Readers are advised to consult Altman (1991: Ch. 5), Armitage and Berry (1987: Ch. 6) or Bland (1987: Ch. 3) for more information, but here is a list of questions (not exhaustive!) which need to be answered before embarking on a medical study:

1 What is the objective of the study?
2 If there are a number of objectives, are there too many?
3 What hypotheses are of interest?
4 Should the design be one of the five types discussed above, or a mixture?
5 If there are alternative designs, what are the pros and cons of each?
6 Which population of patients or subjects do I wish to study?
7 Will I need to take samples from this population?
8 What sample size (study size) will I need, approximately?
9 What variables do I need to measure and/or observe on my patients?
10 How will patients be allocated to the treatments being tested?
11 Do I need a control group of patients?
12 How will the sample data be collected, checked and stored?
13 What statistical analyses will I need to be aware of, even before I collect my data?
14 Should I consult a statistician before I begin my study in earnest?
15 Has somebody else already done a similar study?
16 Are there any ethical issues I need to address?
17 What percentages of patients or subjects of my original sample will:
    (i) refuse to take part;
    (ii) leave the study before it is completed;
    (iii) give rise to missing or erroneous data for one or more variables;
    (iv) not give truthful answers to some questions?

## 14.2 STUDY SIZE

It is unfortunate that so few medical studies discuss how the number of patients or subjects who took part in the study was decided. Often this number seems to have been arrived at by some consideration of administrative convenience – 'we took as many as time and money would allow'. However, using too many patients will almost certainly be a waste of time and money. On the other hand, too few will lead to insufficient **power**, that is, insufficient chance of detecting clinically important differences or effects, assuming they are there to detect (section 8.9). Power is used here as a statistical term, with a particular defined meaning in this context: it is the probability of detecting a clinically important difference, assuming it exists; or equivalently, of accepting the alternative hypothesis, if the alternative hypothesis is true. The latter meaning relates back to the ideas on hypothesis testing of Chapters 8 and so on. We recall there that the significance level, $\alpha$, was the probability of rejecting the null hypothesis if the null hypothesis is true and we conventionally set $\alpha$ at 0.05.

If we now define $\beta$ as the probability of rejecting the alternative hypothesis when the alternative hypothesis is true, then power can be written as $1 - \beta$. It seems obvious that we should make $\beta$ as low as possible, in order to make the power as high as possible.

However, things are not as simple as that because five quantities are inter-connected:

- the study size, $n$;
- the significance level, $\alpha$;
- the power $(1 - \beta)$ to detect a change in;
- $d$, in the parameter we wish to estimate;
- the variability in the variable of interest (in terms of its standard deviation, $\sigma$).

The connection between these five can be expressed in an equation (or graphically in some cases) for given study designs. But the implication of the equation (or graph) is that is we fix any four of the quantities we can determine the fifth. This idea can be used to decide 'objectively' what study size is required.

For the unpaired samples case (section 8.8), where the null hypothesis is $H_0$: $\mu_1 - \mu_2 = 0$, and we wish to detect a difference $d$ between $\mu_1$ and $\mu_2$ in either direction with power $1 - \beta$, the formula is:

$$n = \frac{4(z_{\alpha/2} + z_\beta)^2 \sigma^2}{d^2}$$

where $\alpha$, $\sigma$ and $n$ are defined as above.

For the paired samples case (section 8.7) where the null hypothesis is $H_0$: $\mu_d = 0$, and we wish to detect a difference of $d$ in either direction with power $1 - \beta$, the formula is:

$$n = \frac{(z_{\alpha/2} + z_\beta)^2 \sigma^2}{d^2}$$

were here $\alpha$ is as before $\sigma$ is the population standard deviation of differences, and $n$ is the number of pairs of subjects.

### 14.2.1 Example from case study 6

#### Unpaired samples

From the *Methods* section data we can infer that:

$$\alpha = 0.05, \qquad 1 - \beta = 0.95, \qquad \sigma = 11.4\,\text{dB}, \qquad d = 10\,\text{dB}$$

Hence, $z_{\alpha/2} = 1.96$, $z_\beta = 1.645$ (Table B.1(b)), and so

$$n = \frac{4(1.96 + 1.645)^2 \times 11.4^2}{10^2} = 67.6$$

So 68 subjects (children) are required for each unpaired samples comparison, or 34 in each sample. Assuming a 10% drop-out, the number in each of the four treatment groups is therefore 37 or 38, giving 149 children in all (as stated in the case study).

#### Paired samples

Again from data in the *Methods* section we can infer that $\alpha = 0.05$, $1 - \beta = 0.95$, $\sigma = 14.25$, and $d = 10\,\text{dB}$. Hence $z_{\alpha/2} = 1.96$, $z_\beta = 1.645$ and so

$$n = \frac{(1.96 + 1.645)^2 \times 14.25^2}{10^2} = 26.4$$

The 'subjects' here were 'ears', so we require 26 pairs of ears, or 26 children, for each of four treatment groups, giving 104 children in all. Since more children were needed for unpaired samples comparisons, the latter determined the actual study size used in case study 6.

### 14.2.2 Problems

There are some problems in using the above rather elegant approach to decide study size. First, the two equations used refer only to the cases of unpaired and paired sample data. For other designs including ANOVA, $\chi^2$ tests, regression and correlation, readers should consult Cohen (1992).

Second, we may not be able to specify all the four quantities ($\alpha$, $\beta$, $\sigma$ and $d$) for our particular design, even if it is the usual unpaired samples design, and hence we cannot calculate $n$, the study size.

Third, and refering again to the unpaired samples design, suppose that we could calculate $n$, but the answer we obtained was inpractically large. What could we do?

I will try to address the second and third problems here by introducing a hypothetical example.

### 14.2.3 A study to compare two drugs designed to reduce blood pressure

Drug A is currently in use and is known to be effective in reducing blood pressure (BP) by an average of 20 mm Hg, but this can vary between $-10$ and 50 for extreme cases ($-10$ implies an increase in BP). Drug A has been a standard treatment for 1000 cases per year for five years. A new drug B, is claimed to be more effective than drug A, and the comparable costs and side effects are claimed to be similar. How shall we decide, as GPs, whether to prescribe drug B instead of drug A from now on? What size study would be convincing, and how would it be conducted?

We presumably would like patients with high BP to consent to take part in the study, and then they would be randomly allocated to drug A or B for the trial period. They should not know, nor should the prescribing doctor, which drug they have been allocated. At the end of the trial period their BP will be taken, and the fall in BP calculated. But what is $n$, the number of patients needed to take part in the study?

We need $\alpha$, $\beta$, $\sigma$ and $d$. You can decide yourself and do the calculation if you are a medical professional. I am merely a statistician, but let us see what I come up with.

The null hypothesis is $H_0$: $\mu_A = \mu_B$, meaning that the mean fall in BP for population of patients taking drug A is the same as the mean fall in BP for the population of patients taking drug B. Also, we know that $\alpha$ is the probability of rejecting the null hypothesis when it is true, which, in this example, is the probability of concluding that the mean falls in BP for the two populations are different when they are, in reality, the same. It seems reasonable to set $\alpha = 0.1$, say, which is fairly high. But the 'error' made would only be to introduce a drug which is neither better nor worse than the current one.

We also know that $\beta$ is the risk of rejecting the alternative hypothesis when it is true, which, in this example, is the risk of concluding that there no difference in the mean falls in BP of the two drugs, when in fact there is a difference. It seems reasonable to set $\beta = 0.05$, giving a power of 0.95, because we want to have the better drug at our disposal.

What about $\sigma$? This stands for the standard deviation of the reduction in BP for both drugs A and B (we will assume that these quantities are equal, given no other infomation). We do know that, for drug A, the reduction in BP varies between $-10$ and $+50$ for a largish sample. Assuming that the reduction is appoximately normally distributed, this range is roughly $6\sigma$, since nearly all of a normal distribution lies within three standard deviations of the mean (section 7.5). Hence $50 - (-10) = 6\sigma$, and $\sigma = 10$.

What about $d$? We are given that the effect of drug A is an average reduction of 20. I will take a stab and guess that a further reduction of only 5 would be good, 10 would be excellent, but too much to hope for, so I chose 6, a 30% improvement. Hence, for $\alpha = 0.1$, $\beta = 0.05$, $\sigma = 10$, $d = 6$, we have:

$$n = \frac{4(z_{\alpha/2} + z_\beta)^2 \sigma^2}{d^2}$$

$$= \frac{4(1.645 + 1.645)^2 10^2}{6^2}$$

$$= 120, \text{ or 60 in each sample.}$$

This does not seem impractically large, since BP is a very quick and cheap variable to measure. But suppose that I was limited to a total of $n = 100$? What are my options? I could

1 increase $\alpha$, or
2 increase $\beta$, or

3 increase $d$, or
4 reduce $\sigma$ (No! This is fixed by the results we already have) or
5 some combination of 1, 2, 3.

You can check the following for yourself:

1 Increasing $\alpha$ to 0.2 gives $n = 95$.
2 Increasing $\beta$ to 0.1 gives $n = 95$.
3 Increasing $d$ to 7 gives $n = 88$.

The last result may seem strange – that a smaller sample is required to detect a larger difference. But imagine that the difference in the effects of the drugs was actually very small. Then we would need a lot of data to detect such a difference, and vice versa.

# 15

## Calculators and computers

### 15.1 INTRODUCTION

Most of the calculations performed in Chapters 7–14 can be done quite easily (and were!) on a pocket calculator. Only the calculations of the multiple regression analysis (section 12.6) were done by computer only. However, the trend is towards greater and greater use of computers, so specially written 'statistical computer packages' exist to take some of the drudgery out of the calculations. The problem is that some packages are either too difficult to use because of badly written manuals, or too easy to use so that, in inexperienced hands, data are subject to endless analyses for which they are quite unsuitable. There is no substitute for careful thought before one goes near the computer!.

Data are normally presented to the computer in the same format, irrespective of the package:

| Case number | Variable 1 | Variable 2 | . . . |
|---|---|---|---|
| 1 | x | x | |
| 2 | x | x | |
| 3 | x | | |
| 4 | | | |
| . | | | |
| . | | | |
| . | | | |

Columns represent variables, such as age, sex, blood pressure; while rows represent cases, which in a medical study will usually be patients. Each x represents a number, which is the value of the variable for a particular case. Non-numerical variables are usually input numerically using a code, for example 1 = male, 2 = female. Missing values are best avoided (!) but are sure to occur with large data sets. One way of denoting a missing value is to give a numerical value which is very unlikely to occur in practice. So for blood pressure a value of 0 could denote a missing value, while 99 could be used as a missing value for the variable number of cigarettes smoked per day.

Two statistical computer packages which can be recommended are MINITAB and SPSS-X. The former is very useful in carrying out basic statistical analyses on relatively small data sets with no missing values, such as the few described in the book. (We cannot often use data from the case studies, since the *raw* data we need of the format shown above are rarely presented in published papers.)

## 15.2 MINITAB EXAMPLES

The purpose of this section is to give the reader some idea of how the statistician communicates with the computer and vice versa, using MINITAB, which is a command-driven rather than a menu-driven package. (It is also possible to use a mouse instead of typing in commands if the user prefers this mode of operation.) The first step is to make sure the package is or has been loaded into the computer, and the package is often accessed by the command MINITAB (←) where ← indicates pressing the RETURN or ENTER button. Each line of commands in the examples below should be entered into the computer by pressing this button.

The screen prompt should be MTB>

### EXAMPLE 15.1

Here is a program to input the ages of 50 subjects (section 7.3), to draw a dot-plot of the data, and to calculate some summary statistics:

```
MINITAB
MTB>OUTFILE 'AGEDATA'
MTB>SET C1
```

*Example 15.1* 197

```
DATA>26 26 27 28 29 30 30 32 32 34
DATA>34 34 34 35 35 35 36 36 36 37
DATA>38 38 38 39 39 40 40 40 40 41
DATA>42 42 43 43 44 44 44 45 46 46
DATA>48 49 53 53 55 57 57 58 58 60
DATA>END
MTB> PRINT C1
MTB> NAME OF C1 'AGE'
MTB> DOTPLOT C1
MTB> DESCRIBE C1
MTB> NOOUTFILE
MTB> STOP
OK,  SPOOL AGEDATA.LIS -AT LASER
```

*Note the following observations:*

- Logging into and out of the computer are not covered.
- The command OUTFILE names a file containing virtually all the MINITAB input commands typed in and the corresponding output up to the command NOOUTFILE. The last line, the SPOOL command, enables this file to be printed on to a laser printer. (Notice the name of the file to be spooled ends with .LIS but does not have quotes (') round it as it did in the OUTFILE command.)
- SET C1 implies a set of numbers in column 1, although the numbers may be typed in rows!
- The command END indicates the end of the data.
- To get a printout of your input data on the screen use the command PRINT.
- The NAME command gives a name to the variable stored in column 1.
- The DOTPLOT command results in a dot-plot!
- The DESCRIBE command give ten summary statistics of which the most useful are:
    N, the number of subjects (called $n$ in section 7.3).
    MEAN, the mean age of the subjects (called $\bar{x}$ in section 7.3).
    STDEV, the standard deviation of the ages of the subjects (called $s$ in section 7.3).
- The command STOP, gets you 'out of MINITAB' and back to the computer's operating system.

The file AGEDATA. LIS is as follows:

```
MTB > set c1
DATA> 26 26 27 28 29 30 30 32 32 34
DATA> 34 34 34 35 35 35 36 36 36 37
DATA> 38 38 38 39 39 40 40 40 40 41
DATA> 42 42 43 43 44 44 44 45 46 46
DATA> 48 49 53 53 55 57 57 58 58 60
DATA> end
MTB > print c1

C1
   26 26 27 28 29 30 30 32 32 34 34 34 34
   35 35 35 36 36 36 37 38 38 38 39 39 40
   40 40 40 41 42 42 43 43 44 44 44 45 46
   46 48 49 53 53 55 57 57 58 58 60

MTB > name c1 'age'
MTB > dotplot c1
```

```
                  :..  . :
  : ...: : ::: .: :: .:: :. :  ..       :  . : :   .
--------+---------+---------+---------+---------+--------- age
   28.0      35.0      42.0      49.0      56.0      63.0
```

```
MTB > describe c1

          N    MEAN   MEDIAN  TRMEAN  STDEV  SEMEAN
age      50   40.52    39.50   40.25   8.96    1.27

         MIN    MAX      Q1      Q3
age    26.00  60.00   34.00   45.25

 MTB > nooutfile
```

### EXAMPLE 15.2

This example shows how to obtain a 95% confidence interval for the mean age of a population using the $t$-distribution and the data from section 8.4, and to test the hypothesis that the population mean age is 45 years (section 8.5).

```
MINITAB
MTB>OUTFILE 'AGETEST'
MTB>SET C1
DATA>    33 36 37 38 39 45 50 55 56
DATA>    END
```

*Example 15.3*                    199

```
MTB>NAME C1 'AGE'
MTB>TINTERVAL 95 C1
MTB>TTEST MU=45 C1
MTB>NOOUTFILE
MTB>STOP
OK, SPOOL AGETEST.LIS -AT LASER.
```

There are two observations to make about this program:

- The command TINTERVAL results in a confidence interval based on the formula: $\bar{x} \pm ts/\sqrt{n}$ (section 8.4).
- The command TTEST results in a hypothesis test, as in section 8.5. Notice in the output below that a $p$-value of 0.29 is stated, indicating that the null hypothesis should *not* be rejected at the 5% level of significance (since $0.29 > 0.05$). Refer to section 8.5, note (v), if necessary.

The file AGETEST. LIS is as follows:

```
MTB > set c1
DATA> 27 33 36 37 38 39 45 50 55 56
DATA> end
MTB > name c1 'age'
MTB > tinterval 95 c1

      N   MEAN  STDEV  SE MEAN  95.0 PERCENT C.I.
age  10  41.60   9.59    3.03  (   34.74,  48.46)

MTB > ttest mu=45 c1

TEST OF MU = 45.00 VS MU N.E. 45.00

      N   MEAN  STDEV  SE MEAN       T    P VALUE
age  10  41.60   9.59    3.03   -1.12       0.29

MTB > nooutfile
```

### EXAMPLE 15.3

This example shows how to carry out a $\chi^2$ (chi-square) test for the difference between two proportions, using the example from case study 2 in section 10.5.

```
MINITAB
MTB>OUTFILE 'CHITEST'
MTB>READ C1 C2
DATA>82   24
DATA>669 343
```

```
DATA>END
MTB>PRINT C1 C2
MTB>CHISQUARE C1 C2
MTB>NOOUTFILE
MTB>STOP
OK, SPOOL CHITEST.LIS -AT LASER
```

Note the following concerning the above program and output below:

- The command READ is useful when more than one column of data is input at the same time, while the command SET allows for one column only to be input.
- The output shows the observed frequencies and the expected frequencies which agree with the values in section 10.5. The value of $\chi^2$ shown below, 5.51, does not agree with the 5.01 obtained in section 10.5. This is because MINITAB does not use Yates's correction. However, both 5.01 and 5.51 are greater than 3.84, so the null hypothesis is rejected in both cases. More generally, the use of Yates's correction is advocated for *any* test of $2 \times 2$ contingency table data, since these are always associated with 1 degree of freedom.

```
MTB > read c1 c2
DATA> 82 24
DATA> 669 343
DATA> end
      2 ROWS READ
MTB > print c1 c2

ROW    C1     C2
  1    82     24
  2   669    343

MTB > chisquare c1 c2

Expected counts are printed below observed counts

           C1       C2    Total
  1        82       24      106
        71.20    34.80

  2       669      343     1012
       679.80   332.20
```

*Example 15.4*                   201

```
Total       751       367      1118
ChiSq = 1.637 + 3.350 +
        0.171 + 0.351 = 5.509
df = 1
MTB > nooutfile
```

### EXAMPLE 15.4

The next example shows how to perform simple linear regression analysis, obtain correlation coefficients and aspects of multiple regression analysis. The data are from case study 4 as set out in section 12.6, parts of which were also used in sections 12.2–12.5.

```
MINTAB
MTB>OUTFILE 'REGS'
MTB>READ C1 C2 C3
DATA>11.2 112  35
DATA>6.7   87 −15
DATA>6.5   99  10
DATA>6.7   96   0
DATA>8.0   92 −10
DATA>8.3  102  20
DATA>7.5   91 −15
DATA>6.7   98  10
DATA>7.1  101   5
DATA>6.6   97   5
DATA>END
MTB>NAME C1 'Y'
MTB>NAME C2 'X1'
MTB>NAME C3 'X2'
MTB>PLOT C1 C2────── see note (i)   below
MTB>REGR C1 1 C2───── see note (ii)  below
MTB>CORR C1 C2─────── see note (iii) below
MTB>REGR C1 1 C3───── see note (iv)  below
MTB>REGR C1 2 C2 C3── see note (v)   below
MTB>NOOUTFILE
MTB>STOP
OK, SPOOL REGS.LIS −AT LASER
```

*Notes*

(i) This command produces the scatter diagram as in section 12.2.

(ii) This command produces the regression equation in section 12.2 and the ANOVA table in section 12.4.

(iii) This command produces the correlation coefficient in section 12.3.

(iv) This command produces the regression of $y$ on $x_2$ and the ANOVA table, see section 12.6.

(v) This command produces the multiple regression equation of $y$ on $x_1$ and $x_2$, and the ANOVA table (section 12.6).

```
MTB > read c1 c2 c3
DATA> 11.2 112 35
DATA> 6.7 87 -15
DATA> 6.5 99 10
DATA> 6.7 96 0
DATA> 8.0 92 -10
DATA> 8.3 102 20
DATA> 7.5 91 -15
DATA> 6.7 98 10
DATA> 7.1 101 5
DATA> 6.6 97 5
DATA> end
      10 ROWS READ
MTB > name c1 'y'
MTB > name c2 'x1'
MTB > name c3 'x2'
MTB > plot c1 c2
```

*Example 15.4* 203

```
    11.2+                                        *
        -
y       -
        -
        -
     9.6+
        -
        -
        -
        -                           *
     8.0+              *
        -
        -          *
        -                          *
        -     *        ***
     6.4+                   *
        -
        +---+---+---+----+----+--x1
       85.0 90.0 95.0 100.0 105.0 110.0
```

MTB > regr c1 1 c2

The regression equation is
y = − 6.34 + 0.142 x1

| Predictor | Coef | Stdev | t-ratio | p |
|---|---|---|---|---|
| Constant | −6.342 | 5.199 | −1.22 | 0.257 |
| x1 | 0.14228 | 0.05320 | 2.67 | 0.028 |

s = 1.104    R-sq = 47.2%    R-sq(adj) = 40.6%

Analysis of Variance

| SOURCE | DF | SS | MS | F | P |
|---|---|---|---|---|---|
| Regression | 1 | 8.714 | 8.714 | 7.15 | 0.028 |
| Error | 8 | 9.747 | 1.218 | | |
| Total | 9 | 18.461 | | | |

Unusual Observations

| Obs. | x1 | y | Fit | Stdev.Fit | Residual | St.Resid |
|---|---|---|---|---|---|---|
| 1 | 112 | 11.200 | 9.593 | 0.847 | 1.607 | 2.27R |

R denotes an obs. with a large st. resid.
MTB > corr c1 c2

```
Correlation of y and x1 = 0.687

MTB > regr c1 1 c3

The regression equation is
y = 7.27 + 0.0570 x2

Predictor        Coef      Stdev     t-ratio        p
Constant       7.2736     0.3916       18.57    0.000
x2             0.05699    0.02515       2.27    0.053

s = 1.186    R-sq = 39.1%    R-sq(adj) = 31.5%

Analysis of Variance

SOURCE           DF         SS         MS       F        p
Regression        1      7.217      7.217    5.14    0.053
Error             8     11.244      1.405
Total             9     18.461

Unusual Observations
Obs.    x2        y     Fit   Stdev.Fit   Residual   St.Resid
  1   35.0   11.200   9.268      0.854      1.932      2.35R

R denotes an obs. with a large st. resid.

MTB > regr c1 2 c2 c3

The regression equation is
y = - 16.1 + 0.245 x1 - 0.0468 x2

Predictor        Coef      Stdev     t-ratio        p
Constant       -16.12     19.96       -0.81    0.446
x1             0.2448     0.2087        1.17    0.279
x2            -0.04681    0.09187      -0.51    0.626

s = 1.159    R-sq = 49.1%    R-sq(adj) = 34.5%

Analysis of Variance

SOURCE           DF         SS         MS       F        p
Regression        2      9.063      4.531    3.38    0.094
Error             7      9.398      1.343
Total             9     18.461

SOURCE           DF     SEQ SS
x1                1      8.714
x2                1      0.348
```

*Example 15.4*     205

```
Unusual Observations
Obs.    x1        y     Fit   Stdev.Fit   Residual   St.Resid
  1    112   11.200   9.651      0.896      1.549       2.11R
```

R denotes an obs. with a large st. resid.

MTB > nooutfile

*The laws of God were, she [FN] held, were the laws of life, and these were ascertainable by careful, and especially by statistical, inquiry.*

# *Appendix A*

# Further reading

The following books are recommended as references for the elaboration of some of the topics covered in this book, and for more advanced statistical methods relevant to medicine.

Altman, D.G. (1991) *Practical Statistics for Medical Research*, Chapman & Hall, London.

Armitage, P. and Berry, G. (1987) *Statistical Methods in Medical Research* (2nd edn), Blackwell Scientific Publications, Oxford.

Bland, M. (1987) *An Introduction to Medical Statistics*, Oxford University Press, Oxford.

Cohen, J. (1992) A power primer. *Psychological Bulletin*, 112(1): 115–9.

Rees, D.G. (1989) *Essential Statistics* (2nd edn), Chapman & Hall, London.

*Appendix B*

# Statistical tables

**Table B1(a)** Normal distribution function

| $z = \dfrac{x - \mu}{\sigma}$ | 0.00 | 0.01 | 0.02 | 0.03 | 0.04 | 0.05 | 0.06 | 0.07 | 0.08 | 0.09 |
|---|---|---|---|---|---|---|---|---|---|---|
| 0.0 | 0.5000 | 0.5040 | 0.5080 | 0.5120 | 0.5160 | 0.5199 | 0.5239 | 0.5279 | 0.5319 | 0.5359 |
| 0.1 | 0.5398 | 0.5438 | 0.5478 | 0.5517 | 0.5557 | 0.5596 | 0.5636 | 0.5675 | 0.5714 | 0.5753 |
| 0.2 | 0.5793 | 0.5832 | 0.5871 | 0.5910 | 0.5948 | 0.5987 | 0.6026 | 0.6064 | 0.6103 | 0.6141 |
| 0.3 | 0.6179 | 0.6217 | 0.6255 | 0.6293 | 0.6331 | 0.6368 | 0.6409 | 0.6443 | 0.6480 | 0.6517 |
| 0.4 | 0.6554 | 0.6591 | 0.6628 | 0.6664 | 0.6700 | 0.6736 | 0.6772 | 0.6808 | 0.6844 | 0.6879 |
| 0.5 | 0.6915 | 0.6950 | 0.6985 | 0.7019 | 0.7054 | 0.7088 | 0.7123 | 0.7157 | 0.7190 | 0.7224 |
| 0.6 | 0.7257 | 0.7291 | 0.7324 | 0.7357 | 0.7389 | 0.7422 | 0.7454 | 0.7486 | 0.7517 | 0.7549 |
| 0.7 | 0.7580 | 0.7611 | 0.7642 | 0.7673 | 0.7704 | 0.7734 | 0.7764 | 0.7794 | 0.7823 | 0.7852 |
| 0.8 | 0.7881 | 0.7910 | 0.7939 | 0.7967 | 0.7995 | 0.8023 | 0.8051 | 0.8078 | 0.8106 | 0.8133 |
| 0.9 | 0.8159 | 0.8186 | 0.8212 | 0.8238 | 0.8264 | 0.8289 | 0.8315 | 0.8340 | 0.8365 | 0.8389 |
| 1.0 | 0.8413 | 0.8438 | 0.8461 | 0.8485 | 0.8508 | 0.8531 | 0.8554 | 0.8577 | 0.8599 | 0.8621 |
| 1.1 | 0.8843 | 0.8665 | 0.8686 | 0.8708 | 0.8729 | 0.8749 | 0.8770 | 0.8790 | 0.8810 | 0.8830 |
| 1.2 | 0.8849 | 0.8869 | 0.8888 | 0.8907 | 0.8925 | 0.8944 | 0.8962 | 0.8980 | 0.8997 | 0.9015 |
| 1.3 | 0.9032 | 0.9049 | 0.9066 | 0.9082 | 0.9099 | 0.9115 | 0.9131 | 0.9147 | 0.9162 | 0.9117 |
| 1.4 | 0.9192 | 0.9207 | 0.9222 | 0.9236 | 0.9251 | 0.9265 | 0.9279 | 0.9292 | 0.9306 | 0.9319 |

| z | 0.00 | 0.01 | 0.02 | 0.03 | 0.04 | 0.05 | 0.06 | 0.07 | 0.08 | 0.09 |
|---|------|------|------|------|------|------|------|------|------|------|
| 2.0 | 0.9772 | 0.9778 | 0.9783 | 0.9788 | 0.9793 | 0.9798 | 0.9803 | 0.9808 | 0.9812 | 0.9817 |
| 2.1 | 0.9821 | 0.9826 | 0.9830 | 0.9834 | 0.9838 | 0.9842 | 0.9846 | 0.9850 | 0.9854 | 0.9857 |
| 2.2 | 0.9861 | 0.9864 | 0.9868 | 0.9871 | 0.9875 | 0.9878 | 0.9881 | 0.9884 | 0.9887 | 0.9890 |
| 2.3 | 0.9893 | 0.9896 | 0.9898 | 0.9901 | 0.9904 | 0.9906 | 0.9909 | 0.9911 | 0.9913 | 0.9916 |
| 2.4 | 0.9918 | 0.9920 | 0.9922 | 0.9925 | 0.9927 | 0.9929 | 0.9931 | 0.9932 | 0.9934 | 0.9936 |
| 2.5 | 0.9938 | 0.9940 | 0.9941 | 0.9943 | 0.9945 | 0.9946 | 0.9948 | 0.9949 | 0.9951 | 0.9952 |
| 2.6 | 0.9953 | 0.9955 | 0.9956 | 0.9957 | 0.9959 | 0.9960 | 0.9961 | 0.9962 | 0.9963 | 0.9964 |
| 2.7 | 0.9965 | 0.9966 | 0.9967 | 0.9968 | 0.9969 | 0.9970 | 0.9971 | 0.9972 | 0.9973 | 0.9974 |
| 2.8 | 0.9974 | 0.9975 | 0.9976 | 0.9977 | 0.9977 | 0.9978 | 0.9979 | 0.9979 | 0.9980 | 0.9981 |
| 2.9 | 0.9981 | 0.9982 | 0.9982 | 0.9983 | 0.9984 | 0.9984 | 0.9985 | 0.9985 | 0.9986 | 0.9986 |
| 3.0 | 0.9987 | 0.9987 | 0.9987 | 0.9988 | 0.9988 | 0.9989 | 0.9989 | 0.9989 | 0.9990 | 0.9990 |
| 3.1 | 0.9990 | 0.9991 | 0.9991 | 0.9991 | 0.9992 | 0.9992 | 0.9992 | 0.9992 | 0.9993 | 0.9993 |
| 3.2 | 0.9993 | 0.9993 | 0.9994 | 0.9994 | 0.9994 | 0.9994 | 0.9994 | 0.9995 | 0.9995 | 0.9995 |
| 3.3 | 0.9995 | 0.9995 | 0.9995 | 0.9996 | 0.9996 | 0.9996 | 0.9996 | 0.9996 | 0.9996 | 0.9997 |
| 3.4 | 0.9997 | 0.9997 | 0.9997 | 0.9997 | 0.9997 | 0.9997 | 0.9997 | 0.9997 | 0.9997 | 0.9998 |

For a normal distribution with a mean, $\mu$, and standard deviation, $\sigma$, and a particular value of $x$, calculate $z = (x - \mu)/\sigma$. The table gives the area to the left of $x$, see figure below.

0.8944

$$\frac{(x - \mu)}{\sigma} \quad \frac{15 - 10}{\ } = 1.25$$

**Table B1(b)** Upper percentage points for the normal distribution

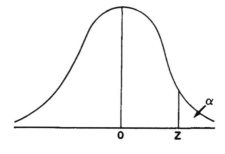

| $\alpha$ | 0.05 | 0.025 | 0.01 | 0.005 | 0.001 | 0.0005 |
|---|---|---|---|---|---|---|
| z | 1.645 | 1.96 | 2.33 | 2.58 | 3.09 | 3.29 |

The table gives the value of z for various right-hand tail areas, $\alpha$.

**Table B.2** Upper percentage points for the *t*-distribution

| $\alpha =$ | | 0.05 | 0.025 | 0.01 | 0.005 | 0.001 | 0.0005 |
|---|---|---|---|---|---|---|---|
| $v =$ | 1 | 6.31 | 12.71 | 31.82 | 63.66 | 318.3 | 636.6 |
| | 2 | 2.92 | 4.30 | 6.96 | 9.92 | 22.33 | 31.60 |
| | 3 | 2.35 | 3.18 | 4.54 | 5.84 | 10.21 | 12.92 |
| | 4 | 2.13 | 2.78 | 3.75 | 4.60 | 7.17 | 8.61 |
| | 5 | 2.02 | 2.57 | 3.36 | 4.03 | 5.89 | 6.87 |
| | 6 | 1.94 | 2.45 | 3.14 | 3.71 | 5.21 | 5.96 |
| | 7 | 1.89 | 2.36 | 3.00 | 3.50 | 4.79 | 5.41 |
| | 8 | 1.86 | 2.31 | 2.90 | 3.36 | 4.50 | 5.04 |
| | 9 | 1.83 | 2.26 | 2.82 | 3.25 | 4.30 | 4.78 |
| | 10 | 1.81 | 2.23 | 2.76 | 3.17 | 4.14 | 4.59 |
| | 12 | 1.78 | 2.18 | 2.68 | 3.05 | 3.93 | 4.32 |
| | 14 | 1.76 | 2.14 | 2.62 | 2.98 | 3.79 | 4.14 |
| | 16 | 1.75 | 2.12 | 2.58 | 2.92 | 3.69 | 4.01 |
| | 18 | 1.73 | 2.10 | 2.55 | 2.88 | 3.61 | 3.92 |
| | 20 | 1.72 | 2.09 | 2.53 | 2.85 | 3.55 | 3.85 |
| | 25 | 1.71 | 2.06 | 2.48 | 2.79 | 3.45 | 3.72 |
| | 30 | 1.70 | 2.04 | 2.46 | 2.75 | 3.39 | 3.65 |
| | 40 | 1.68 | 2.02 | 2.42 | 2.70 | 3.31 | 3.55 |
| | 60 | 1.67 | 2.00 | 2.39 | 2.66 | 3.23 | 3.46 |
| | 120 | 1.66 | 1.98 | 2.36 | 2.62 | 3.16 | 3.37 |
| | $\infty$ | 1.64 | 1.96 | 2.33 | 2.58 | 3.09 | 3.29 |

The tabulated value is $t_{\alpha,v}$, where $p\ (x > t_{\alpha,v}) = \alpha$, when $x$ has the $t$ distribution with $v$ degrees of freedom.

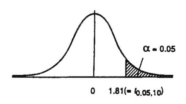

$\alpha = 0.05$

0    1.81(= $t_{0.05,10}$)

*t* distribution with $v = 10$df

**Table B.3** 5 per cent points of the $F$-distribution

| $v_1 =$ | 1 | 2 | 3 | 4 | 5 | 6 | 7 | 8 | 10 | 12 | 24 | $\infty$ |
|---|---|---|---|---|---|---|---|---|---|---|---|---|
| $v_2 = 1$ | 161.4 | 199.5 | 215.7 | 224.6 | 230.2 | 234.0 | 236.8 | 238.9 | 241.9 | 243.9 | 249.1 | 254.3 |
| 2 | 18.5 | 19.0 | 19.2 | 19.2 | 19.3 | 19.3 | 19.4 | 19.4 | 19.4 | 19.4 | 19.5 | 19.5 |
| 3 | 10.1 | 9.55 | 9.28 | 9.12 | 9.01 | 8.94 | 8.89 | 8.85 | 8.79 | 8.74 | 8.64 | 8.53 |
| 4 | 7.71 | 6.94 | 6.59 | 6.39 | 6.26 | 6.16 | 6.09 | 6.04 | 5.96 | 5.91 | 5.77 | 5.63 |
| 5 | 6.61 | 5.79 | 5.41 | 5.19 | 5.05 | 4.95 | 4.88 | 4.82 | 4.74 | 4.68 | 4.53 | 4.36 |
| 6 | 5.99 | 5.14 | 4.76 | 4.53 | 4.39 | 4.28 | 4.21 | 4.15 | 4.06 | 4.00 | 3.84 | 3.67 |
| 7 | 5.59 | 4.74 | 4.35 | 4.12 | 3.97 | 3.87 | 3.79 | 3.73 | 3.64 | 3.57 | 3.41 | 3.23 |
| 8 | 5.32 | 4.46 | 4.07 | 3.84 | 3.69 | 3.58 | 3.50 | 3.44 | 3.35 | 3.28 | 3.12 | 2.93 |
| 9 | 5.12 | 4.26 | 3.86 | 3.63 | 3.48 | 3.37 | 3.29 | 3.23 | 3.14 | 3.07 | 2.90 | 2.71 |
| 10 | 4.96 | 4.10 | 3.71 | 3.48 | 3.33 | 3.22 | 3.14 | 3.07 | 2.98 | 2.91 | 2.74 | 2.54 |
| 12 | 4.75 | 3.89 | 3.49 | 3.26 | 3.11 | 3.00 | 2.91 | 2.85 | 2.75 | 2.69 | 2.51 | 2.30 |
| 15 | 4.54 | 3.68 | 3.29 | 3.06 | 2.90 | 2.79 | 2.71 | 2.64 | 2.54 | 2.48 | 2.29 | 2.07 |
| 20 | 4.35 | 3.49 | 3.10 | 2.87 | 2.71 | 2.60 | 2.51 | 2.45 | 2.35 | 2.28 | 2.08 | 1.84 |
| 24 | 4.26 | 3.40 | 3.01 | 2.78 | 2.62 | 2.51 | 2.42 | 2.36 | 2.25 | 2.18 | 1.98 | 1.73 |
| 30 | 4.17 | 3.32 | 2.92 | 2.69 | 2.53 | 2.42 | 2.33 | 2.27 | 2.16 | 2.09 | 1.89 | 1.62 |
| 40 | 4.08 | 3.23 | 2.84 | 2.61 | 2.45 | 2.34 | 2.25 | 2.18 | 2.08 | 2.00 | 1.79 | 1.51 |
| 60 | 4.00 | 3.15 | 2.76 | 2.53 | 2.37 | 2.25 | 2.17 | 2.10 | 1.99 | 1.92 | 1.70 | 1.39 |

**Table B.4** Critical values of studentized range statistic, $q$

| | | | | $a = 0.05$ | | | | | |
|---|---|---|---|---|---|---|---|---|---|
| $a(n-1)$ | $p = 2$ | 3 | 4 | 5 | 6 | 7 | 8 | 9 | 10 |
| 1 | 17.97 | 26.98 | 32.82 | 37.08 | 40.41 | 43.12 | 45.40 | 47.36 | 49.07 |
| 2 | 6.085 | 8.331 | 9.798 | 10.88 | 11.74 | 12.44 | 13.03 | 13.54 | 13.99 |
| 3 | 4.501 | 5.910 | 6.825 | 7.502 | 8.037 | 8.478 | 8.853 | 9.177 | 9.462 |
| 4 | 3.927 | 5.040 | 5.757 | 6.287 | 6.707 | 7.053 | 7.347 | 7.602 | 7.826 |
| 5 | 3.635 | 4.602 | 5.218 | 5.673 | 6.033 | 6.330 | 6.582 | 6.802 | 6.995 |
| 6 | 3.461 | 4.339 | 4.896 | 5.305 | 5.628 | 5.895 | 6.122 | 6.319 | 6.493 |
| 7 | 3.344 | 4.165 | 4.681 | 5.060 | 5.359 | 5.606 | 5.815 | 5.998 | 6.158 |
| 8 | 3.261 | 4.041 | 4.529 | 4.886 | 5.167 | 5.399 | 5.597 | 5.767 | 5.918 |
| 9 | 3.199 | 3.949 | 4.415 | 4.756 | 5.024 | 5.244 | 5.432 | 5.595 | 5.739 |
| 10 | 3.151 | 3.877 | 4.327 | 4.654 | 4.912 | 5.124 | 5.305 | 5.461 | 5.599 |
| 11 | 3.113 | 3.820 | 4.256 | 4.574 | 4.823 | 5.028 | 5.202 | 5.353 | 5.487 |
| 12 | 3.082 | 3.773 | 4.199 | 4.508 | 4.751 | 4.950 | 5.119 | 5.265 | 5.395 |
| 13 | 3.055 | 3.735 | 4.151 | 4.453 | 4.690 | 4.885 | 5.049 | 5.192 | 5.318 |
| 14 | 3.033 | 3.702 | 4.111 | 4.407 | 4.639 | 4.829 | 4.990 | 5.131 | 5.254 |
| 15 | 3.014 | 3.674 | 4.076 | 4.367 | 4.595 | 4.782 | 4.940 | 5.077 | 5.198 |
| 16 | 2.998 | 3.649 | 4.046 | 4.333 | 4.557 | 4.741 | 4.897 | 5.031 | 5.150 |
| 17 | 2.984 | 3.628 | 4.020 | 4.303 | 4.524 | 4.705 | 4.858 | 4.991 | 5.108 |
| 18 | 2.971 | 3.609 | 3.997 | 4.277 | 4.495 | 4.673 | 4.824 | 4.956 | 5.071 |
| 19 | 2.960 | 3.593 | 3.977 | 4.252 | 4.469 | 4.645 | 4.794 | 4.924 | 5.038 |
| 20 | 2.950 | 3.578 | 3.958 | 4.232 | 4.445 | 4.620 | 4.768 | 4.896 | 5.008 |
| 24 | 2.919 | 3.532 | 3.901 | 4.166 | 4.373 | 4.541 | 4.684 | 4.807 | 4.915 |
| 30 | 2.888 | 3.486 | 3.845 | 4.102 | 4.302 | 4.464 | 4.602 | 4.720 | 4.824 |
| 40 | 2.858 | 3.442 | 3.791 | 4.039 | 4.232 | 4.389 | 4.521 | 4.635 | 4.735 |
| 60 | 2.829 | 3.399 | 3.737 | 3.977 | 4.163 | 4.314 | 4.441 | 4.550 | 4.646 |
| | 2.800 | 3.356 | 3.685 | 3.917 | 4.096 | 4.241 | 4.363 | 4.468 | 4.560 |
| ∞ | 2.772 | 3.314 | 3.633 | 3.858 | 4.030 | 4.170 | 4.286 | 4.387 | 4.474 |

able may be used in a posterior test following the analysis of variance for a completely randomized design, with factor $A$ at $a$
s and $n$ observations at each level. (Here $p$ is the difference in the ranks of a pair of treatment means +1.)

**Table B.5** Percentage points for the $\chi^2$ distribution

| $\alpha =$ | 0.999 | 0.995 | 0.99 | 0.975 | 0.95 | 0.9 | 0.1 | 0.05 | 0.025 | 0.01 | 0.005 | 0.001 |
|---|---|---|---|---|---|---|---|---|---|---|---|---|
| $\nu = 1$ | $0.0^5157^*$ | $0.0^4393$ | $0.0^3157$ | $0.0^3982$ | $0.0^2393$ | 0.0158 | 2.71 | 3.84 | 5.02 | 6.63 | 7.88 | 10.83 |
| 2 | $0.0^2200$ | 0.0100 | 0.0201 | 0.0506 | 0.103 | 0.211 | 4.61 | 5.99 | 7.38 | 9.21 | 10.60 | 13.81 |
| 3 | 0.0243 | 0.0717 | 0.115 | 0.216 | 0.352 | 0.584 | 6.25 | 7.81 | 9.35 | 11.34 | 12.84 | 16.27 |
| 4 | 0.0908 | 0.207 | 0.297 | 0.484 | 0.711 | 1.06 | 7.78 | 9.49 | 11.14 | 13.28 | 14.86 | 18.47 |
| 5 | 0.210 | 0.412 | 0.554 | 0.831 | 1.15 | 1.61 | 9.24 | 11.07 | 12.83 | 15.09 | 16.75 | 20.52 |
| 6 | 0.381 | 0.676 | 0.872 | 1.24 | 1.64 | 2.20 | 10.64 | 12.59 | 14.45 | 16.81 | 18.55 | 22.46 |
| 7 | 0.599 | 0.989 | 1.24 | 1.69 | 2.17 | 2.83 | 12.02 | 14.07 | 16.01 | 18.48 | 20.28 | 24.32 |
| 8 | 0.857 | 1.34 | 1.65 | 2.18 | 2.73 | 3.49 | 13.36 | 15.51 | 17.53 | 20.09 | 21.95 | 26.12 |
| 9 | 1.15 | 1.73 | 2.09 | 2.70 | 3.32 | 4.17 | 14.68 | 16.92 | 19.02 | 21.67 | 23.59 | 27.88 |
| 10 | 1.48 | 2.16 | 2.56 | 3.25 | 3.94 | 4.86 | 15.99 | 18.31 | 20.48 | 23.21 | 25.19 | 29.59 |
| 12 | 2.21 | 3.07 | 3.57 | 4.40 | 5.23 | 6.30 | 18.55 | 21.03 | 23.34 | 26.22 | 28.30 | 32.91 |
| 14 | 3.04 | 4.08 | 4.66 | 5.63 | 6.57 | 7.79 | 21.06 | 23.68 | 26.12 | 29.14 | 31.32 | 36.12 |
| 16 | 3.94 | 5.14 | 5.81 | 6.91 | 7.96 | 9.31 | 23.54 | 26.30 | 28.85 | 32.00 | 34.27 | 39.25 |
| 18 | 4.90 | 6.26 | 7.02 | 8.23 | 9.39 | 10.86 | 25.99 | 28.87 | 31.53 | 34.81 | 37.16 | 42.31 |
| 20 | 5.92 | 7.43 | 8.26 | 9.59 | 10.85 | 12.44 | 28.41 | 31.41 | 34.17 | 37.57 | 40.00 | 45.31 |
| 25 | 8.65 | 10.52 | 11.52 | 13.12 | 14.61 | 16.47 | 34.38 | 37.65 | 40.65 | 44.31 | 46.93 | 52.62 |
| 30 | 11.59 | 13.79 | 14.95 | 16.79 | 18.49 | 20.60 | 40.26 | 43.77 | 46.98 | 50.89 | 53.67 | 59.70 |
| 40 | 17.92 | 20.71 | 22.16 | 24.43 | 26.51 | 29.05 | 51.81 | 55.76 | 59.34 | 63.69 | 66.77 | 73.40 |

# Index

Printed and bound by CPI Group (UK) Ltd, Croydon, CR0 4YY

22/10/2024

01777605-0003